ALSO BY HAL H. HARRISON

A Field Guide to Birds' Nests

A Field Guide to Western Birds' Nests

The World of the Snake

American Birds in Color

Outdoor Adventures

Pennsylvania Birdlife

Nesting Birds of Sanibel-Captiva

Pennsylvania Reptiles and Amphibians

Wood Warblers' World

BY
HAL H. HARRISON

*Technical Assistance and Range Maps
by Mada Harrison*

SIMON AND SCHUSTER NEW YORK

Published by Simon and Schuster
A Division of Simon & Schuster, Inc.
Simon & Schuster Building
Rockefeller Center
1230 Avenue of the Americas
New York, New York 10020
SIMON AND SCHUSTER and colophon are registered
trademarks of Simon & Schuster, Inc.
Designed by Eve Kirch
Manufactured in the United States of America
1 3 5 7 9 10 8 6 4 2
Library of Congress Cataloging in Publication Data
Harrison, Hal H.
Wood warblers' world.
Includes index.
1. Wood warblers. I. Title.
QL696.P2618H37 1984 598.8'72 84-1335
ISBN 0-671-47798-6

All photographs were taken by the author except the following:

COLOR PHOTOGRAPHS

Arthur A. Allen: Plate 18 (bottom).
Emmet R. Blake: Plate 2 (top right).
Bob and Elsie Boggs: Plate 8 (bottom).

(Continued on page 335)

For my grandchildren,
Pete and Jennie
and
Scotty and Griff

Acknowledgments

The idea and inspiration for a book about Wood Warblers illustrated with photographs of birds in the wild and written principally from research and experience gleaned through years of field work came to me about thirty years ago. In the meantime, I was sidetracked from my original idea by assignments to produce two Peterson Field Guides to birds' nests and eggs. Always, however, well established in the back of my mind was the determination to complete the warbler book and fulfill my ambition.

To segregate the many friends who assisted me in producing the warbler book from those who helped me with my field guides would be impossible. To a great extent, they are the same people; but there are some who deserve special thanks and I acknowledge them here.

As always, my wife, Mada, has been my constant companion in the field and in our office. Her expertise in locating breeding territories, studying behavior, and making important notes has improved season by season. In the office she is the typist, editor, and always-available consultant.

My son, George, another Simon & Schuster author, is also my literary agent. This may be the only such father-son relationship extant. Both George and his wife, Kit, have been helpful in reviewing my work.

Before publication, the entire text of *Wood Warblers' World* was read critically by Dr. Kenneth C. Parkes, chief curator of Life Sciences and curator of Birds at the Carnegie Museum of Natural History in Pittsburgh. His critical comments were vital to the accuracy of the text.

I am grateful to others who read individual chapters before publication; they contributed much to the high standards set for the book. Their expertise regarding certain species added to the validity of many chapters. My thanks go to Emmet R. Blake, Betty Darling Cottrille, Steven W. Eaton, Millicent S. Ficken, George A. Hall, Brooke Meanley, Howard L. Mendall, Harold F. Mayfield, Val Nolan, Jr., Warren M. Pulich, Jay Shuler, Robert M. Stewart, and Lawrence H. Walkinshaw.

Throughout the years devoted to research for the book, no field companions were more helpful to me than Joe Grom and Ken Vierling in Pennsylvania and Ralph Long in Maine. Hardly a year passed that we were not together in fields, forests, and swamps.

During the spring and summer in the Chiricahua Mountains in Arizona and two seasons on Mount Desert Island, Maine, I was fortunate to have A. Blake Gardner as a field assistant. His tree-climbing ability is matched only by his never-ending enthusiasm for field work and photography.

Jay Shuler, of McClellanville, South Carolina, did everything possible to find for me the rare Bachman's Warbler. It was not Jay's fault that we failed.

Because of his personal interest and the high quality of his workmanship, Ray Quigley of Whittier, California, is deserving of my thanks for superior work in his darkroom studio.

Janet Hinshaw, librarian for the Wilson Ornithological Society at the University of Michigan, was always helpful in supplying research material, much of which was unavailable to me elsewhere.

Many personal friends helped me, but none was more enthusiastic, consistently encouraging, and determined that the warbler book would be published than Betty and Powell Cottrille of Jackson, Michigan. Their many letters, faithfully written during the past few years, were always an inspiration. And to add to my good fortune, Betty is one of the best bird photographers I know.

In Betty's class as an expert in bird photography is Bill Dyer, who winters in Florida and summers in Michigan and who takes great warbler pictures in both places. A number of his best photographs contribute greatly to the beauty of the book.

Special thanks must go to Dan Johnson, my editor at Simon & Schuster. Dan not only is an expert editor but is a most patient person with determined writers like me.

The list is long and I know I will be remiss in leaving out some valuable friends, but I can remember with thanks the following who provided information from research, assistance in field work, opportunities for photography, and encouragement in my endeavors: Edith Andrews, John Arvin, W. Wilson Baker, Griffing Bancroft, James Bond, Mary Lou Brown, Dale Dalrymple, John Dennis, Greg E. Eddy, Marguerite Geibel, Mary Gray, Ruth Grom, Ed Harrison, Grant Heilman, Julia Janosik, Faith Kaltenbach, Lloyd Kiff, Ed Komarek, Roy Komarek, H. R. Leapman, Robert C. Leberman, Les Line, Barry Mansell, Janet Mathison, Ted and Lois Matthews, Russell T. Norris, Vernon Norris, Joe Panza, Roger Tory Peterson, Frank W. Preston, Russell L. Pyke, Ruth and Warner Reeser, Carl and Laura Richter, Eleanor Sims, Merit Skaggs, Bart Snyder, Henry M. Stevenson, John Strohm, George M. Sutton, William H. Thompson, George E. Watson, and Chen W. Young.

Contents

Color plates of all species appear following pages 64, 160 and 256.

Foreword

Wassaic Creek is born in a soggy depression in a cow pasture about two miles above our upstate New York retreat. In its youth, it is little more than an open ditch, sloshing across a muddy barnyard, through the stubble of a cornfield. Paralleling the country road, gathering run-off from fields to the left and right, the creek falls 350 feet in those two miles, and during day-long rains it becomes a brown torrent, seven steps wide, by the time it reaches the bridge to my wife's garden.

Here Wassaic Creek makes a sharp turn away from the road and enters, if only briefly for there is another cow pasture just downstream, a whole new world. We are a two-bridge family, and the second bridge spans the stream where those occasional torrents have cut a gash thirty feet deep through a limestone cliff. Here the creek, rippling clear on sunny days, is shaded by ash and birch and maple and locust. Here a spring tumbles over mossy rocks, past a curtain of maidenhair fern, into an eddy above a very modest rapids. Here, on the first weekend in May, I can count on finding the Northern Waterthrush.

I should be candid at this point and confess to an affinity for the nonconformists in the Wood Warbler clan. "Aberrant species," Roger Tory Peterson calls them in his field guide. The Ovenbird, the Waterthrushes, the Yellow-breasted Chat. Warblers that fail to behave like warblers. Perhaps I am also aberrant, but I would prefer one raucous pair of chats in a tangled, thorny fencerow to a forest full of *Dendroicas.* That is why I will spend an entire spring morning trying to spy the Ovenbird, whose call is so loud and whose habits are so secretive. That is why I will ignore the colorful pageantry in the greening canopy if I find a waterthrush teetering by my creek, snatching at fishlings and aquatic insects and worms, its lovely liquid song ringing over the burble of falling water.

I do not know if the Northern Waterthrush nests on my "2.3 acres more or less," as the deed reads. My postage-stamp woodland is cool and damp, there is abundant foodstuff in and near the creek, and there are appropriate mossy niches and upturned tree roots where eggs could be hidden. I suspect the Northern Waterthrush does live here. But, as Hal Harrison notes in his eloquent and informative essay on the species, one can watch this elusive warbler for hours without finding its nest. And I am less inclined today to pry into the lives of such creatures than I was in my formative birding years, when banding records and breeding bird surveys and a burgeoning life-list were so important. I realize that a nesting waterthrush would add spice to a rather pedestrian roster of resident birds. But I am content with the knowledge that I can count on its visit each spring.

An Ovenbird I do not have. I must hike a neighbor's woods—an honest-to-God forest up the mountainside—to hear its *teacher, teacher, TEACHER!* Nor a Yellow-breasted Chat, for we are at the very northern edge of its range. As for other more customary Wood Warblers, a fair share of those species, which one would expect in these near-New England hills, pause in my treetops to refuel during spring migration waves. I do not ignore them, the Black-throated Blue, Myrtle, Magnolia, Canada, Chestnut-sided, Redstart, or Yellowthroat. They are, as Hal Harrison writes, "among the most beautifully colored and distinctively patterned of all birds."

In the popular literature on North American birdlife, the warblers have been given short shrift. There are many books on waterfowl, on gamebirds, on raptors, even a handful on shorebirds. *Wood Warblers' World,* then, fills a real need and does so with both literary élan and thoroughness, plus presenting splendid photography as a bonus. Next May, when I sit on my downstream bridge waiting for the appearance of the Northern Waterthrush in my ferny little gorge, I will bring along my copy and read about its conformist cousins flitting overhead.

—Les Line
Editor, *Audubon*

Introduction

American Warblers . . . exclusively our own . . . throughout the world we find no finer group of birds, thus they may well be considered the pride of the American ornithologist.

Those words, written over eighty years ago by Charles Johnson Maynard, are echoed today by several million American bird watchers who delight at the sight of Wood Warblers, called by Roger Tory Peterson "the butterflies of the bird world."

Through my long association with bird watchers, bird clubs, and even those only casually interested in birds, I have learned that warblers are first—or at least very high—on everyone's list of favorite birds. This is understandable. They are among the most beautifully colored and distinctively patterned of all birds, and there is fascinating variety in their behavior. This great interest among bird watchers is indicative of their desire to know more about the lives of the fifty-three Wood Warblers that nest in the United States and Canada. The purpose of this book is to satisfy that curiosity in a manner that is accurate, informative, interesting, and entertaining.

During my more than thirty years of studying warblers in their summer homes and during migrations north and south (and, in many cases, associating with them in their winter homes), I have made notes of interesting aspects of the Wood Warblers' world. In addition, I have researched the experiences of other observers to add to my own knowledge.

I assume that anyone who reads this book owns a field guide to identification, so I have not discussed in detail field marks that are known to millions of birders. Nests and eggs are mentioned generally,

Wood Warblers, like this Black-throated Blue male, are high on everyone's list of favorite birds. Note the slender bill and grasping feet—three toes forward and one toe back.

but most details have been omitted since field guides to these subjects are also available.

As most bird watchers know only too well, many of the male migratory Wood Warblers that nest north of the tropics molt in late summer and become what are popularly known as "confusing fall warblers." Identification in autumn is one of the most challenging facets of the sport of bird watching. Drab females and immatures compound the confusion, and here, too, field guides are helpful in sorting out species.

Since readers are likely to have the opportunity to visit warblers on their breeding grounds, maps show where each species may be expected to nest. Winter ranges are outlined in the text.

The order in which the species are discussed, the English names, and the scientific names follow the dictates of the sixth edition of the American Ornithologists' Union *Check-list of North American Birds,* published in 1983.

WHAT IS A WOOD WARBLER?

Early American naturalists, most of whom came here from Europe, encountered for the first time a family of birds whose members are found only in the Western Hemisphere. Most of these pioneers mistook the birds for Old World Warblers, or Sylvia, a family to which our kinglets and gnatcatchers belong (now known as Sylviidae). Between 1835 and 1840 it became clear that the species found in America were different. The word *warbler,* however, was well established; and the accepted name became American Wood Warblers to distinguish them from the European family. The word *wood* was accepted because most (though not all) members of the family are woodland residents.

Actually, the word *warbler* is a misnomer for the American group. *The Random House Dictionary* defines *warble* as "to sing with trills" and indicates that the word is derived from the Middle English word *werble,* meaning a tune. Most of our warblers do not warble. The males all sing, but not always well. They lisp, buzz, hiss, chip, rollick, or zip; and one, the Yellow-breasted Chat, may chuckle, hoot, whistle, caw, and screech.

The American Ornithological Union Check-list Committee in 1983 placed Wood Warblers in the subfamily Parulinae (par-you-LIE-nee) in the family Emberizidae (em-ber-EYE-zid-ee). They are referred to as parulines (PARR-you-lines). Previously Wood Warblers were in a family of their own, Parulidae.

Wood Warblers are perching birds (Passeriformes). This means that they are considered to be among the birds most recently evolved; and they are still evolving rapidly. The feet of all perching birds have three toes forward and one turned back with which the bird can grip a perch (branch, twig, wire, grass stem). Their leg muscles are formed so that the grip on the perch tightens if the bird sways backward.

Warblers are small, mainly insect-eating birds with slender sharp-pointed bills. The bills are somewhat flattened in some species, but are never hooked like those of vireos, which are often confused with warblers. Warblers are generally more active than vireos. Bird banders who handle thousands of warblers annually are aware that they have nine primaries, the largest flight feathers in the wings. This is a distinguishing characteristic of Wood Warblers and others of their family of birds; most other Passeriformes have ten primaries.

It is correct to refer to warblers as songbirds because they are Oscines, a division of Passeriformes that are commonly called songbirds. Most of the world's best singers belong to this group; however, I consider Wood Warblers generally to be mediocre songsters at best.

A warbler's characteristic nine primary flight feathers are distinguishable only in the hand. This Black-and-white Warbler has been mist-netted for banding by Barbara Patterson, Somesville, Maine.

This doesn't mean they don't do a lot of singing, and their musical efforts are part of the fun of warbler watching and listening. Describing what warblers "say" as they sing is not only difficult but sometimes impossible (e.g., try imitating the Yellow-breasted Chat). It is surprising how many warbler songs observers have compared to the chipping rattle of the Chipping Sparrow (about fifteen percent of all species).

When warblers return to us in the spring, the males usually arrive first. Females follow soon after. Each male promptly stakes out a nesting territory and proclaims ownership by patrolling its boundaries and by singing almost constantly. In accepting her mate, the female accepts his territory and builds her nest there. Wood Warbler courtship is not, in most cases, a spectacular event. Song seems to be the principal occupation of the courting male. A few species are known to engage in a flight song rendered at various heights, mostly above treetops. Physical contact between mated birds appears to be confined to the few seconds required for copulation.

Nests are built almost invariably by the female, and nest sites range from hollows in the ground to the tops of the tallest trees. The number of eggs varies in general from three to seven in a single clutch, with four or five being the average. All are white or nearly white and all are spotted and blotched with shades of brown, lilac, or black except the plain white eggs laid by Bachman's and Swainson's Warblers. Commonly the markings are concentrated or wreathed at the larger end of the egg. Incubation by the female may take from eleven to fourteen days, rarely longer. Young remain in the nest from eight to eleven days. Normally both parents feed the young. Although warblers are hatched blind and with sparse down, in a few weeks they are ready to fly wherever their migration route takes them.

As a group, warblers are plagued by the parasitic Brown-headed Cowbird possibly more than any other group of birds. In addition to my own records, I have quoted freely from the work of Dr. Herbert Friedmann, who has made a life study of cowbirds and has been the official compiler of cowbird records for many years.

A male Palm Warbler sings his weak, repetitious *zee-zee-zee-zee-zee* notes while patrolling his territory.

This nest of a Chestnut-sided Warbler holds three Brown-headed Cowbird eggs (right) and three warbler eggs.

In addition to the fifty-three species I have reviewed, there are many Wood Warblers that spend their entire lives in the tropics—Mexico, Central America, South America, and the West Indies. Among nonmigratory warblers of the tropics, the sexes are generally alike in plumage, even those whose feathers are most colorful. The Painted Redstart and Red-faced Warbler, two primarily Mexican species that do reach the southwestern United States as nesting birds, are examples. The male and female Black-throated Blue Warblers show the greatest difference in plumage among the more northern species.

MIGRATION

The vast majority of Wood Warblers leave their summer homes in Canada and the United States in midsummer or early fall to winter in Mexico, Central America, South America, and the West Indies.

Why do they migrate? The theory that birds move south to escape cold northern winters is only a partial answer. Many birds, even those as tiny and frail-looking as chickadees, reside in the northern United States all year.

The probable answer may surprise you—birds migrate to go home. They are not "our" birds. It has been suggested that the migrant species that come to us from the tropics in spring originated in the tropics and have evolved as migrants in order to improve their chances of breeding successfully by lessening pressures from other breeding birds and from predators.

The eminent ornithologist Dr. Paul Schwartz put it this way: "Migrant species should not be considered 'invaders' to the tropics but as species that have tropical niches the same as resident species. During two-thirds of the year, including the periods of least resources for insectivorous birds, the neotropics harbors and has harbored for thousands of years an avifauna that is best treated as a single unit. Some of the species of the avifaunal unit depart for a brief period each year to reproduce. They return to the tropics not as invaders with a disturbing impact but to their long-ago won places."[1]

Many of our Wood Warblers average only about three months in their breeding grounds, two to three months in migration, and six to seven months on their wintering grounds. In fact, some Yellow Warblers leave their breeding areas as early as the first two weeks in July; by early August, practically all are gone. In Canada's Churchill area, this species arrives, nests, and has departed within a period of seven weeks; Yellow-rumped Warblers, Blackpoll Warblers, and Northern Waterthrushes take about ten weeks.

Most species of warblers go south of the United States in migration, but the distance they travel varies. Scarcely any remain throughout the winter in the area where they nested. Among the exceptions are the races of the Common Yellowthroat that are permanent residents in the southeasternmost and southwesternmost United States. Another is one subspecies of the Orange-crowned Warbler, a permanent resident on the Santa Barbara Islands, California. The Pine Warbler is somewhat migratory but travels no farther south than the southern part of its breeding range. In fact, the Pine is unique among warblers; in winter it is confined almost entirely to the United States. Almost all warblers of the western United States spend the winter in Mexico or Central America.

[1] Schwartz, Paul, "Some Considerations on Migratory Birds," in *Migrant Birds in the Neotropics,* edited by Allen Keast and Eugene S. Morton (Washington, D.C.: Smithsonian Institution Press, 1980), p. 31.

This Yellow Warbler may have built a nest, incubated eggs, fledged young, and started back to its home in the tropics by mid-July.

The Pine is unique among warblers; its winter range is confined almost entirely to southern United States.

Although warblers occasionally migrate by day, most are night fliers. The long trip of five hundred to seven hundred miles across the Gulf of Mexico is evidently made nonstop in a single night's flight. One of the many marvels of migration is the fact that a tiny feathered creature weighing less than a half-ounce, with loose feathers hardly adapted to withstand strong winds and heavy rains, can cross and recross the Gulf of Mexico not only once, but sometimes for several years. This dangerous crossing takes a heavy toll of birds annually.

Birds are not entirely unprepared for the arduous journey in spring and fall. While migration is synchronized to the seasons, in addition to responding to a stimulus, each bird must be ready to meet the energy requirements of a long flight. Energy is stored by eating food in excess of daily requirements, thereby accumulating fat.

Unquestionably birds migrate in enormous numbers. Dr. Sidney A. Gauthreaux, Jr., of Clemson University, surveyed migrating birds on the night of September 28, 1977, at the Greenville-Spartanburg airport in South Carolina, with radar and telescope. Scanning a line across the path of migration, he computed the number of birds that passed through in a six-hour period at more than a million. In one hour he got a peak count of 218,700.[2]

Unlike most migrants, warblers move north together in mixed groups. Flocks may be large or small, but during the height of migration it is unusual to find a flock made up of a single species, although one species may predominate.

Because they are basically insect-eating birds, warblers are generally late-spring migrants. Earlier migrants, like the Black-and-white Warbler, move slowly across land in spring migration, averaging about thirteen miles per day. Late migrants move faster. The Canada Warbler, for example, averages thirty miles per day in its journey from the Gulf of Mexico to Canada.

Many hazards await migrating birds and mortality is great. Probably no other group suffers more than the parulines. Among those hazards are man-made perils such as lighthouses, TV towers, and tall buildings, and natural disasters such as drought, floods, fire, sudden cold snaps, snow, and gales.

[2] Fisher, Allen C., Jr. "Mysteries of Bird Migration," *National Geographic Magazine,* vol. 156, no. 2 (1979), p. 165.

Among the many hazards confronting warblers on their long migrant journeys spring and fall are TV towers. Warblers, nocturnal migrants, fly into these towers and are killed.

Studies of nocturnal disasters have been made in various parts of the country, and all indicate that there is a tremendous loss of life because migrants fly into towers at night. One example is a TV tower in Orange County, Florida, erected in 1969 by Station WDBO-TV. Daily records of birds killed were kept for August, September, October, November, and December in 1969, 1970, and 1971. Sixteen families and eighty-two species were represented in the 7,782 individuals collected during that period. Yellowthroats (2,710), Black-throated Blue Warblers (856), Ovenbirds (714), American Redstarts (579), and

Palm Warblers (517) made up 69 percent of the total. Thirty-eight per cent of the 82 species and 85 percent of the 7,782 individuals were warblers. At the WFMS-TV tower in Youngstown, Ohio, as many as 150 Ovenbirds have been killed in a single night.

Despite the fact that weather undoubtedly plays an important part in departure and progress of migration, there is nevertheless a regularity over a period of years in the arrival time of warblers in spring and departure time in fall. Those who keep annual records of spring arrivals can predict the date within a few days. Sometime in mid-May is the "big day" for bird watchers in the northern states and Canadian provinces. One of the biggest days in the history of birdwatching at Point Pelee National Park, Ontario, was May 11, 1963, when 143 species of birds were counted. Trees were filled with warblers, hundreds and hundreds of them.

One of the great pleasures in the sport of birdwatching is to see the great waves of migrant warblers that pass through northern woods in spring.

Many bird watchers who visit Point Pelee in May are apt to visit Cape May, New Jersey, between September 15 and 25 to watch the spectacle in reverse. Warblers and other migrants heading south along the Atlantic Coast funnel into Cape May Point, which is a takeoff spot for an over-water flight. Some will cross Delaware Bay and proceed down the coast; others will continue southward over the Atlantic Ocean. A count of between 100 and 150 different species in a few days during this period is not uncommon.

WINTER

When migrant warblers, which have spent the winter in the tropics, leave for their northern breeding grounds in the spring, vast numbers of other nonmigrant (resident) bird species and individuals remain. Winter residents in the tropics, which leave each spring to nest in the Temperate Zone, spend six months or more in their migration flights and on their breeding grounds. One might assume that integrating with the resident birds after returning in the fall would be a hazardous

Winter range of the endangered Kirtland's Warbler is limited to the scattered Bahama Islands, a vast area of about forty-five hundred square miles.

undertaking. This is not the case. Much research has been done throughout the tropics, and all authorities agree that the resident tropical birds do not fill the niches left by the departing migrants in spring. So how do these migrants interrelate with the resident parulines and with other species that share their habitats in the tropics? John T. Emlen, of the Department of Zoology, University of Wisconsin, who has studied wintering parulines primarily in Florida and the Bahama pinelands, offers this explanation: "Winter integration is considered to have evolved by selective processes of displacement and extinction; few traces of unresolved competition persist. Food shortage is probably not a critical factor in population regulation for most species. . . ."[3]

Studying ecological aspects of migrant bird behavior in Veracruz, Mexico, John H. Rappole and Dwain W. Warner of the University of Minnesota conclude: "Most migrant species in our study were territorial and showed long term site fidelity to their wintering areas, indicating that these species were able to compete with permanent residents for resources in primary forests and elsewhere.[4]

Although few species remain close to their nesting areas in winter, not all Wood Warblers winter south of the United States. Undoubtedly the hardiest is the Yellow-rumped Warbler. In winter, the eastern race of this species is found along the Atlantic Coast from New England to Florida and along the Gulf Coast westward through Texas and north through the Mississippi Valley. The western race winters in the southwestern United States and south into Central America. Both the Palm and Orange-crowned Warblers have ranges similar to the Yellow-rumped but do not remain as far north in winter. The Common Yellowthroat, and the Prairie and Yellow-throated Warblers are year-round residents in the Southeast, principally in Florida. Except for the Pine Warbler and the eastern race of the Yellow-rumped Warbler, most individuals of the species mentioned here do leave the United States in winter.

The National Audubon Society's annual Christmas counts regularly bring surprises, for throughout North America warblers turn up in most unlikely places: Yellow-breasted Chats in Nova Scotia; a Yellow

[3] Emlen, John T., "Interaction of Migrants and Resident Land Birds in Florida and Bahama Pinelands," in Keast and Morton, eds., p. 133.

[4] Ibid. p. 389.

Warbler in Ontario; a Black-throated Blue Warbler at Northampton, Massachusetts; a Cape May Warbler wintering at a feeding station in Pittsburgh; a Chestnut-sided Warbler at Tucson Valley, Arizona; four Palm Warblers at Tillamook Bay, Oregon; and others.

Princeton biologist John W. Terborgh reveals some interesting facts about the areas south of the United States occupied by wintering migrants. In terms of absolute numbers, it is clear that border and offshore areas are preferred to more distant ones. Terborgh writes: "In Mexico and the Bahamas, migrants commonly make up 50 percent or more of the winter populations in a broad spectrum of both natural and disturbed habitats. Progressively more remote destinations in the Caribbean harbor fewer and fewer migrants. . . . A similar pattern holds along the Central American isthmus and into South America. . . . It seems likely that as much as half of all the land birds that go south of the United States each winter funnel into Mexico, the Bahamas, Cuba, and Hispaniola, which offer a combined area of

Many wintering warblers are attracted to feeding stations that offer fresh fruit. In Florida, cut-open oranges are especially attractive.

This atypical Cape May Warbler spent the winter at a feeding station near Pittsburgh, Pennsylvania. It ate peanut butter, suet, and cornmeal.

2,175,000 square kilometers as compared to 16,200,000 square kilometers for North America south of the tree line. Rough as these numbers are, they strongly imply that many migratory populations are concentrated severalfold on their wintering grounds."[5]

CONSERVATION

As I write this in Fort Lauderdale, Florida, in the spring of 1983, I am concerned about the rapacious destruction of brushy areas of a county park that for years has been my favorite local birding spot. Broward County is a playground for tourists and displaced northerners, and therein lies the reason for so-called improvement and modernization. Until now, Markham Park has offered to campers and local citizens only birds, trees, flowers, ferns, cattails, woodland trails, picnicking, a bit of fishing in the canals, even solitude. Much of this is being wiped out in order to establish tennis courts, a lake, bridle paths, swimming pools, target ranges for gun lovers, a zoo, and probably other features yet unannounced.

[5] Terborgh, John W., "The Conservation Status of Neotropical Migrants: Present and Future," in Keast and Morton, eds., p. 22.

"Progress" comes to a wild Florida park. This was the home of Prairie Warblers, Common Yellowthroats, White-eyed Vireos, Loggerhead Shrikes, towhees and other resident birds. Wintering here were great flocks of Palm and Yellow-rumped Warblers.

Every winter the park has been alive with warblers—Palm, Yellow-rumped, Black-and-white, Common Yellowthroat, Prairie, and occasional visitors such as Ovenbirds, Orange-crowned, and Yellow-throated Warblers. What will happen to these birds and the many other species that winter or live permanently in the park?

It is erroneous to believe that birds displaced by shopping centers, housing developments, golf courses, swamp drainage, and similar works of man will simply move to a new and undisturbed location. The "new and undisturbed" locations already have their full quota of birds. Observers who are conscious of the slow but steady decline of many species know that more than any other factor, loss of habitat is the cause.

It is not only the loss of habitat in our own country that has the conservationists worried. In recent years, ecologists have become increasingly alarmed over the continual loss of habitat in tropical America for migrants from the Temperate Zone. Highland forests in Mexico, Central America, and northern South America are rapidly being preempted for agriculture, fuel, and lumber.

It is alarming to note that twenty-nine species of Wood Warblers winter in mature tropical forests. John W. Terborgh points out, "Clearing one hectare of forest in Mexico is equivalent to expanding urban sprawl by perhaps 5–8 hectares in the Northeast.

Twenty-nine species of Wood Warblers winter in mature tropical forests like this one in Colombia, South America. Continued deforestation will result in major reductions in their numbers.

"With over half the natural vegetation of Central America and the Greater Antilles already converted to cropland and pasture, and the remainder disappearing at a rate of a few per cent a year, we seriously face the prospect that suitable habitat will no longer be available for many migrant species by the end of the century."[6]

Terborgh sounds this warning: "Continued deforestation in the near neotropics will result in major reductions in the numbers of many forest-dwelling migrants. We are, in effect, about to play observers in a massive experiment in which there will be dramatic alterations in the relative population sizes of numerous common species.

[6] Ibid.

A monument commemorating efforts to preserve the nesting range of Kirtland's Warbler was erected in Mio, Michigan, on July 27, 1963, by citizens of Oscoda County. This is the world's first monument honoring a songbird.

No one can yet say which species will be most affected, or what all the consequences will be."[7]

We are told that none of our common species are likely to become extinct, but numbers of the more specialized species will decrease and their places will be taken by more aggressive and adaptable birds. According to Terborgh, "There are already signs that this is taking place in some fragmented eastern woodlands. . . . The total biomass of birds breeding in the North American continent will probably change little. What will change is the familiar ambience of our forests in springtime. It just won't sound the way it used to."[8]

[7] Ibid., p. 29.
[8] Ibid.

In the United States a "conservation alert" is conducted by the National Audubon Society in its coverage of bird populations throughout the United States and Canada. Changes in population of any species become evident through the cooperative work of many teams who census migrants, wintering birds, and breeding populations. Their findings are published in *American Birds,* a bi-monthly journal published by the National Audubon Society. This early alert is a definite warning signal; but just what can be done when the population of any species is found to be declining is problematic.

Wood Warblers benefit from the protection given all birds in local, state, and national refuges, parks and sanctuaries; but they continue to be subject to the hazards of pollution and continuing development of land. As far as I know, Kirtland's is the only species of Wood Warbler that has been singled out for special conservation attention. In the counties in Michigan where Kirtland's Warbler breeds, a special Kirtland's Warbler Management Area has been set up. (Details of this effort are given in chapter 27.) To date, Kirtland's and Bachman's are the only warblers that have received endangered species status from the Fish and Wildlife Service of the United States Department of the Interior.

Interest in proper management of forest habitats for birds was significantly evidenced recently when the Forest Service, United States Department of Agriculture, published the book *Forest Habitat for Birds of the Northeast.* In the foreword, Steve Yurich, Regional Forester, makes the following statement:

"This publication provides habitat requirements for birds associated with a forest environment. The information will assist federal and state land management agencies, as well as private individuals, to plan for the habitat needs of birds in the management of public and private forest lands."[9]

This seems to be a step in the right direction, and we can hope and plan for more steps to follow.

[9] DeGraaf, Richard M., et al., *Forest Habitat for Birds of the Northeast* (Washington D.C.: Forest Service, United States Department of Agriculture, 1980), p. iii.

1. *Bachman's Warbler*

Vermivora bachmanii

PLATE 2

Folks around Awendaw and McClellanville on Route 17 north of Charleston, South Carolina, instantly recognize the bird watchers who invade the cypress and palmetto swamps in their Carolina low country each spring. It isn't only binoculars and Peterson Field Guides that set them apart. It's also their declaration that they have come to the Francis Marion National Forest to hunt for one of America's rarest birds, BOCKman's Warbler. The natives know that the Bachman family prefers to be called BACKman.

The cypress swamps are beautiful and a spring visit is memorable, but if the success of your journey depends upon adding Bachman's to your life list, you are facing tremendous odds. Since the early 1960s, when a number of sightings were substantiated by authorities, only a few isolated reports have been received from South Carolina, Georgia, and Louisiana. Despite the odds, Bachman's Warbler, an endangered species, remains the prime challenge to bird watchers. In 1978, the American Birding Association (ABA) asked its members to list the birds most wanted for their life lists, and Bachman's Warbler was number one, ahead of both the California Condor and the Ivory-billed Woodpecker.

During the past thirty years, I have made several forays into the I'On Swamp, principally in the Fairlawn Plantation in the vicinity of Myrant's Reserve, near McClellanville. Fairlawn is a privately owned part of the swamp where Arthur T. Wayne found thirty-two nests of this species between 1906 and 1918. (Fairlawn is fenced and the gates are locked. Permission is needed before entering.) No nest has been seen in South Carolina since the last one Wayne found on April 11, 1918.

I'On Swamp near Charleston, South Carolina, where Arthur T. Wayne found thirty-two nests of the rare and endangered Bachman's Warbler before 1918. No nests have been found here since.

Under the guidance of Jay Shuler, who has spent years documenting the lives of Audubon, Bachman, Wayne, and the elusive warbler, I last visited I'On Swamp in March 1979. During a week of intense searching, we heard or saw nothing resembling Bachman's Warbler, even though the swamp was alive with birds, some migrants, some permanent residents, all announcing the arrival of spring.

Bachman's Warbler has never been common. Few persons have seen it since its discovery by Reverend John Bachman in July 1832, near Parker's Ferry on the Edisto River about thirty-five miles west of Charleston. In the few sightings reported, the male has been seen most often. Bachman first collected a female; then in March 1833, he collected a second specimen, a male. When he turned the two skins over to his close friend, John James Audubon, the great naturalist named the bird for its discoverer. He called it Bachman's Swamp Warbler. The warbler was not seen again in the United States for fifty years.

Over sixty years later the first nest of the warbler was found by Otto Widmann on May 17, 1897, on Kolb Island, Dunklin County, Missouri. Widmann found a second nest on May 13, 1898, on Buffalo

Island in Dunklin County. George C. Embody found a nest with three eggs in Kentucky in 1906, and two nests have been found in Alabama. Most bird specimens in museum collections were taken along the birds' migration route, especially on Key West, Florida.

The bird's rarity is only one factor contributing to the paucity of our information regarding its life history. Wayne, who discovered the first South Carolina nest in 1906, had the greatest opportunity to study Bachman's nest life; but he was a professional egg collector who could sell a set of Bachman's eggs to other amateur or professional collectors for about one hundred fifty dollars even in those early days, and wasn't interested in studying birds.

Henry M. Stevenson, one of the finders of an Alabama nest in 1937, states, "For some reason, Bachman's Warbler seems to be poorly equipped for survival even under the conditions it demands and is not

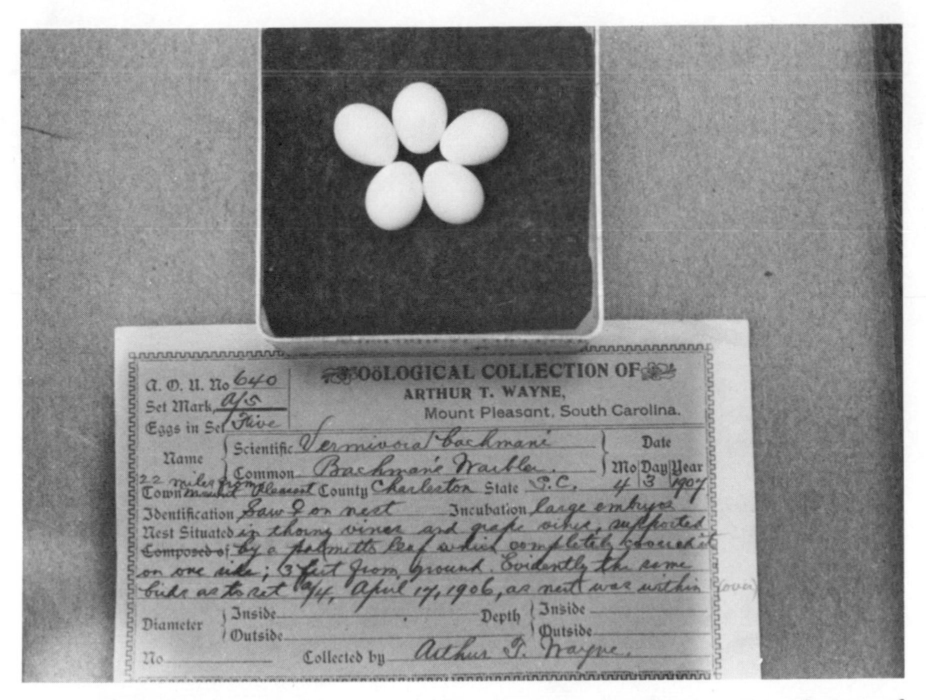

A set of Bachman's Warbler eggs, like these in the Western Foundation of Vertebrate Zoology's collection, sold for $150 in 1906. These were collected April 3, 1907, by Arthur T. Wayne according to the data sheet shown with them.

sufficiently adaptable to survive under different conditions. In the long process of evolution, flourishing, and eventual extinction of species, perhaps this is one whose time has come."[1]

Jay Shuler cites heavy lumbering and the resulting destruction of swamp habitat to which Bachman's Warbler is attached as a major factor in the species' decline. Shuler maintains that "no animal or plant, no matter how well adapted, can long survive if the habitat for which it evolved is abruptly destroyed."[2]

Some forestry specialists contend that lumbering does not affect the nesting habitat of Bachman's Warbler. The rationale for this belief is the fact that the birds nest in low, bushy cover and not in trees.

Intrepid birders determined to search for this bird and its nest are advised that they may encounter some unexpected hardships. Practically all nests found by Wayne were concealed in dense undergrowth usually near or over water. Searchers hacked their way through thick bushes and dense tangles of catbrier, cane, and scrub palmetto. Ticks, chiggers, and mosquitoes made life miserable and there was always the danger of encountering rattlesnakes and cottonmouths.

Males feed and sing in the upper branches of large swamp trees. They usually remain motionless while they sing, making them very difficult to see. The song is distinctive but hard to describe. Wayne likened it to songs of the Worm-eating Warbler and the Northern Parula. Thomas A. Imhof writes: "The song is a buzzy trill, all in one pitch but sometimes with a higher introductory note, and so ventriloquial as to make the singer difficult to locate. Although the song resembles those of the Parula and Worm-eating Warblers, its buzzing quality strongly suggests the kinship of Bachman's Warbler to the Golden-winged and Blue-winged Warblers."[3]

Bachman's eggs are porcelain white with a translucence that lets the yolk impart a pink blush in the early stages of incubation. Swainson's Warbler, a close swamp neighbor of Bachman's, is the only other Wood Warbler that lays pure white unmarked eggs. I'On Swamp re-

[1] Stevenson, Henry M., "The Recent History of Bachman's Warbler," Wilson Bulletin, vol. 84, no. 3, p. 347.

[2] Personal correspondence.

[3] Imhof, Thomas A., *Alabama Birds* (Tuscaloosa, Alabama: University of Alabama Press, 1962), p. 445.

cords indicate that four eggs is the average clutch for Bachman's Warbler, but three and five have been found.

In 1980, five observers reported a female Bachman's Warbler in Cuba. We know the species winters in Cuba and on the Isle of Pines, migrating south through Florida and the keys as early as late July. In spring it returns north as early as late February. Bachman's Warblers have been seen building nests in mid-March.

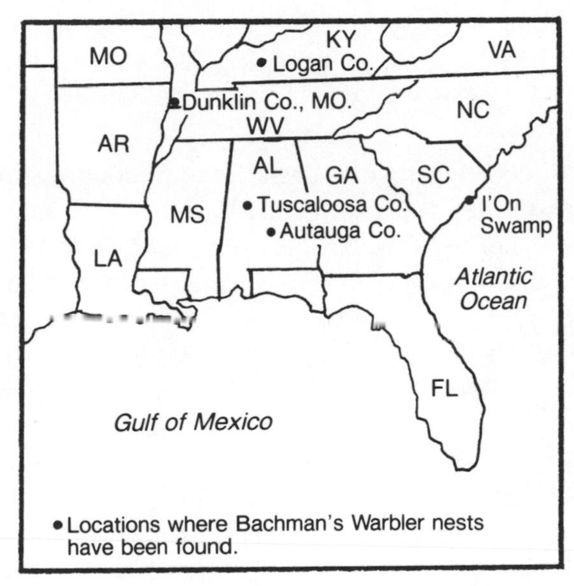

Breeding Range of Bachman's Warbler

2. *Blue-winged Warbler*

Vermivora pinus

PLATE 1

Every spring since I began years ago to photograph nesting birds, the bane of my existence has been predation. Chipmunks and squirrels are my number one suspects, but there are many others—snakes, jays, raccoons, weasels, skunks. It seems to me that ground-nesting birds suffer more than others simply because of their ready availability to so many predators. Of this I am certain: I have lost more Blue-winged Warbler nests to predators than those of any other ground-nesting warbler.

My experiences in 1983 are fresh in my mind. During May and June, I had available to me six Blue-winged Warbler nests. When found, three held young and three contained eggs. All were destroyed, and in no case did human interference account for the loss. I have learned long ago the folly of disturbing a nest that I want to photograph at a later time. In one case, my wife flushed a female Blue-winged from her nest in a patch of goldenrod. Mada cautiously peered into the nest. It held four eggs. She immediately reported to me, and within ten minutes I was at the site. The nest was empty.

A beginning bird watcher can be forgiven for thinking this brightly colored warbler uncommon during migration. However, when Blue-wingeds have settled in their preferred habitat of overgrown pastures, woodland edges, bottomlands, swamps, and edges of small streams, the males can usually be located without much difficulty. The male may expose himself when he mounts to a bush or tree to sing, but the female is not likely to be seen unless glimpsed near her nest.

The male's song can easily be ignored if the bird watcher is not aware that it is more insectlike than birdlike. It has been described as

Woodland edges, abandoned fields grown up to thick shrubs, stream edges, and bottomlands are desirable habitats for Blue-winged Warblers.

A Blue-winged Warbler displays the large white patches on its outer tail feathers when the tail is fanned.

The nest of a Blue-winged Warbler is a bulky structure, large for the size of the female that builds it. The coarse material of the outer walls blends with its surroundings.

a two-parted locustlike *zwe-e-e-e-e-e ze-e-e-e-e-e,* the latter part a mere sputter and usually lower in pitch and rougher in sound than the first. The song just described is the Blue-winged Warbler's territorial song, common for the first week or two after the males arrive on their breeding grounds. Later, the nesting song begins. It has the same buzzing quality but is longer and variable, not always given twice alike.

Blue-winged Warblers leave their winter homes in southern Mexico or northern Central America (casual in Panama and rare in the West Indies) in March. They make the demanding journey across the Gulf of Mexico and arrive on their nesting grounds in late April or early May. Nesting begins by late May and early June.

To find a nest, I have learned, requires time and patience, for it is hidden close to or on the ground. The female is generally reported as building the nest alone; however, I watched a male bring material as

The Blue winged Warbler female is a close sitter and sometimes allows the searcher to touch her before she flushes.

often as the female during the first day of construction at a site in Butler County, Pennsylvania. For her size, the female builds a bulky nest. It is hidden among and attached to upright stems of a clump of forbs or grass. The nest is deep and sometimes narrow and is supported on a sturdy foundation of dry leaves.

When surprised on her nest, the female slips off quietly into surrounding vegetation without any attempt to lure the intruder away. She is a very close sitter and sometimes allows the searcher to touch her before she flushes.

Dr. Friedmann considers the Blue-winged Warbler a fairly frequent victim of the Brown-headed Cowbird. He has records of at least fifty-two cases of parasitization, of which sixteen were from New York State.

It is my experience that in Pennsylvania, where I have found the most Blue-winged Warbler nests, this species builds a nest very similar to that of its close relative, the Golden-winged Warbler, but tends to prefer more moist conditions. In fact, I found one nest buried in a clump of skunk cabbage in a wet swamp. However, like the Golden-winged, it also nests in old clearings and abandoned overgrown fields. A common nesting site for either species is deep in a clump of goldenrod.

The Blue-winged and Golden-winged Warblers hybridize frequently, and the continual increase in range overlap brought about by the Blue-winged's expansion of range has resulted in a greater frequency of hybrid forms: Brewster's and Lawrence's Warblers. Both hybrids were once thought to be full species; the former was named *Vermivora leucobronchialis* and Lawrence's was called *Vermivora lawrencei.* The suggestion that *Vermivora pinus* and *Vermivora chrysoptera* be considered conspecific because of the interbreeding has been rejected by most modern ornithologists. It appears the Golden-winged and Blue-winged Warblers are too distinctly different to be lumped. While the hybrids are definitely fertile, they very rarely, if

This male Brewster's Warbler, a hybrid offspring of a Golden-winged and Blue-winged interbreeding, was mated to a female Golden-winged Warbler at this nest in Michigan.

Young Blue-winged Warblers can flutter away from the nest when eight days old. Undisturbed, they remain for about ten days. This youngster has fledged.

ever, breed with each other but instead are almost always recorded as breeding with either of the parental types.

In Passaic County, New Jersey, T. Donald Carter observed a nesting Brewster's Warbler for six years. In each of five years that his nest was discovered, the hybrid was found to be mated to a female Golden-winged Warbler. Through observation and banding it was determined that the Brewster's had a different mate each year.

The more common Brewster's Warbler is thought to be the offspring of the original cross between a Blue-winged Warbler and a Golden-winged Warbler. The rarer Lawrence's Warbler represents a combination of recessive characteristics and could result from a number of different crosses among the hybrid forms. For a scholarly and detailed discussion of this avian phenomenon, see Kenneth C. Parkes's article, "The Genetics of the Golden-winged × Blue-winged Warbler Complex," listed in the bibliography.

Neither the generic name, *Vermivora* (ver-MIV-oh-rah), meaning worm-eating, nor the species name, *pinus,* Latin for pine tree, is appropriate. It is believed that *pinus* was derived from "pine creeper," the name by which the Blue-winged was known to the early historians of American ornithology Mark Catesby and George Edwards. Another early name for the species was Blue-winged Yellow Warbler.

Edwards and Catesby were responsible for many of the earliest descriptions of American birds. Edwards was an English colonist who explored the American colonies between 1741 and 1751 and compiled a record called *History of Uncommon Birds.* Catesby, also English and a contemporary of Edwards, was an author and illustrator who worked chiefly in the Carolinas.

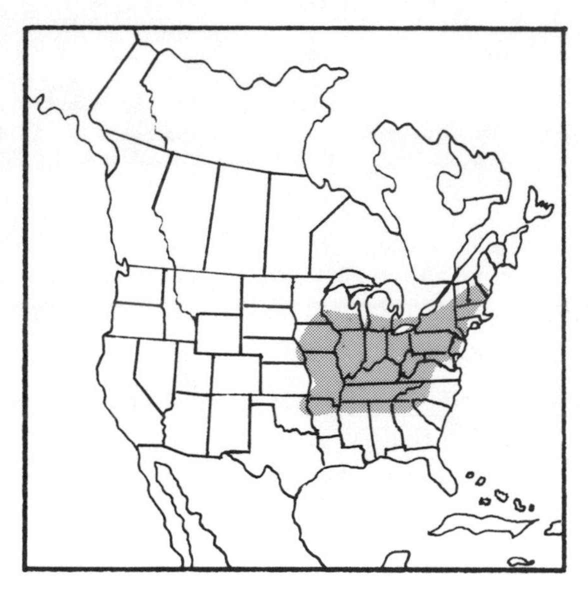

Breeding Range of Blue-winged Warbler

3. *Golden-winged Warbler*

Vermivora chrysoptera

PLATE 3

Yes, I have been aware that I hear fewer Golden-winged Warblers now than I did twenty or thirty years ago in my field work in western Pennsylvania; but no, I did not consider this anything but a local problem. However, recent studies have indicated that this condition is widespread, that the Golden-winged Warbler population is declining, and conversely the population of Blue-winged Warblers is increasing. The two conditions are apparently related.

We have observed that the population of any species may fluctuate from year to year. Indeed, some populations, like those of the Ruffed Grouse, are normally cyclic. But when the number of breeding pairs of a Wood Warbler continues to decline in areas where historically it has been common, it is time for concern.

The Blue-winged and Golden-winged Warblers are principals in the best-known, most intensively studied instances of hybridization between bird species in North America. Offspring are known as Brewster's and Lawrence's Warblers. (See the discussion of this phenomenon in the preceding chapter.)

It has become evident to ornithologists in recent years that where the breeding ranges of the Blue-winged and Golden-winged Warblers overlap for a number of years, the population of the latter declines. Frank B. Gill, who has studied the results of this establishment of sympatry, writes: "At present I know of no localities with breeding Golden-wings where Blue-wings have been established for fifty years or more."[1]

[1] Gill, Frank B., "Historical Aspects of Hybridization Between Blue-winged and Golden-winged Warblers," *The Auk,* vol. 97, no. 1 (1980), p. 14.

Populations of Golden-winged Warblers are declining, while Blue-winged Warblers are increasing. The sight of a female feeding her young at a typical nest on a foundation of leaves may become rare.

Theories explaining this situation are varied. It is possible that Blue-wingeds actually outcompete in meeting the demand for food and nesting sites and for that reason replace Golden-wingeds. On the other hand, we have learned that each species of bird demands a particular type of habitat for nesting and has specific preferences with respect to kind of soil, density or type of vegetation, and other aspects of its surroundings. Based on this understanding, we may wonder if the Golden-winged Warbler has very special requirements and is not adapting to changing land practices such as the abandonment of large tracts of farmland, which may mean a loss of suitable habitat. The thriving Blue-winged Warbler seems to be more accepting and willing or able to accommodate to such change.

Gill makes this ominous prediction: "If the present trend continues for another 100 years, it seems probable that the Golden-winged Warbler will be a very rare species, if not extinct."[2]

[2]Ibid., p. 15.

Hillside thickets, openings in deciduous forests, and woodland edges offer ideal nest sites for Golden-winged Warblers.

This species winters in Central America and south to Colombia and Venezuela and, rarely, in the West Indies. It arrives on its breeding grounds in late April or early May. Like the Blue-winged Warbler, it has been extending its range northward and eastward in a broad area east of the Mississippi. Apparently range expansion of the Blue-winged has been more rapid and the Golden-winged is free of competition from it only in the extreme northern part of its range and in higher elevations of the Appalachian Mountains.

While the habitats preferred for nesting by the Blue-winged and Golden-winged Warblers overlap to some extent, I find that the Golden-winged generally demands higher, drier ground. From their studies in Tompkins County, New York, in 1978, biologists John L. Confer and Kristine Knapp conclude that the Blue-winged Warbler has a tolerance for a wide range of habitats but that the Golden-winged Warbler nests in a habitat that exists for a relatively brief time during succession of plants from abandoned farmlands to brushy fields. They conclude that the Blue-winged is a habitat generalist, the Golden-winged a specialist.

Eggs of the Golden-winged Warbler are similar to those of the Blue-winged Warbler but usually are more heavily marked.

The song of the male Golden-winged Warbler is an insect-like series of buzzy notes: *beeeee-buzz-buzz-buzz.* This male has just fed young at the nest.

The nest of the Golden-winged is not unlike that of the Blue-winged Warbler: on or close to the ground, supported by weed stalks (often goldenrod) or tufts of grass. The four or five eggs are generally more heavily marked than the similar eggs of the Blue-winged. The nest also might be mistaken for that of the Common Yellowthroat, which also builds a bulky nest on or near the ground, often in a clump of goldenrod, although the Golden-winged Warbler nest is generally darker in color.

To me, the typical insectlike song of the Golden-winged Warbler is composed of four buzzy notes: *beeeee-buzz-buzz-buzz.* The first note is prolonged and is followed by three shorter notes on a lower pitch. The first *beeeee* note makes up about half of the song, which is drawling in tempo. There is an almost imperceptible pause after the *beeeee* before the final notes are given. Although this is the common song, variations are occasionally heard. Indeed, the Golden-winged sometimes sings the similar song of its close relative, the Blue-winged Warbler, and vice versa; hybrids may sing either parental song, or even both.

Millicent S. Ficken and Robert W. Ficken have observed interaction between the Black-capped Chickadee and Golden-winged

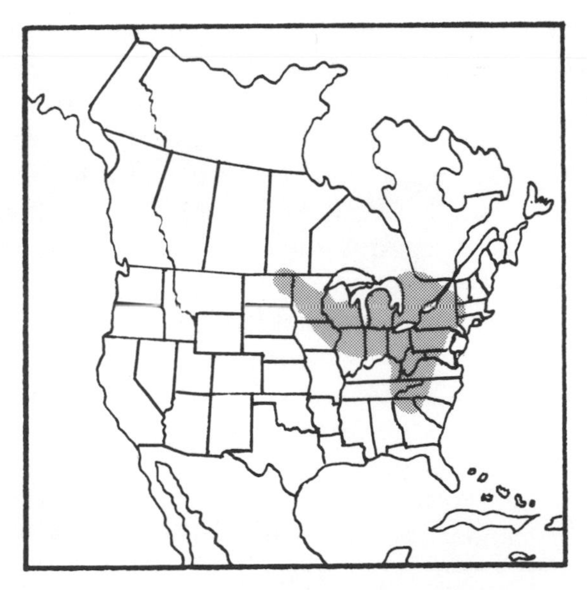

Breeding Range of Golden-winged Warbler

The author and his wife, Mada, photographing the nest of a Golden-winged Warbler at the base of a clump of honeysuckle sprouts in western Pennsylvania.

Warbler. Not only are plumages of the two somewhat similar (gray, white, and black), and the *beeeee-buzz* song of the warbler and the *chicka-dee-dee* call of the chickadee somewhat alike in general pattern, but the two species often move about together with little aggression. Social mimicry may be involved, and both may benefit from this association in some way, according to the Fickens.

That the Golden-winged Warbler is a rather common host to the Brown-headed Cowbird is indicated in records kept by Dr. Friedmann. Added to seventeen instances known before 1963 are nineteen more from Massachusetts, New York, and Ontario, Canada.

Chrysoptera (kris-OP-ter-ah), the specific name of the Golden-winged Warbler, is from the Greek *chrysos,* meaning gold and *pteron,* wing. Karl (or Carl) von Linné (Carolus Linnaeus), a Swedish botanist, early in the 1700s originated the binomial system still used worldwide for naming and classifying plants and animals. Linnaeus himself named the bird. His acceptance of the new species was based on a description recorded by George Edwards.

4. *Tennessee Warbler*

Vermivora peregrina

PLATE 5

How many of us would ever see a Tennessee Warbler if the male didn't sing? He may not sing well but he certainly sings loudly.

This vireolike warbler is so plain and undistinguished that it seldom would be noticed were it not for its loud, persistent notes. The song is somewhat like that of its close relative, the Nashville Warbler, but the Nashville's song is typically two-parted while the Tennessee's is usually three-parted. But not always! I have often heard a Tennessee stop abruptly after the second phrase and occasionally after the first phrase. Each part consists of a repeated note or notes and differs from the other parts in pitch, rhythm, and duration. It is one of the loudest of warbler songs.

The Tennessee Warbler presents a marvelous example of great distances traveled in migration. In spring, it moves northward from its winter home in northern South America, Central America, and Mexico into Texas, where it fans out to its extensive northern breeding range. The journey north is primarily an inland route, centering on the Mississippi Valley. In late summer and early fall, after a few months spent in nesting activities in Canada and the northern United States, this tiny bird, less than five inches long and weighing less than one ounce, returns overland as far as three thousand miles to reach its winter home.

I have always been impressed with the restlessness of the male Tennessee Warbler. Unlike the vireos, with which it is compared, it is anything but deliberate as it feeds or sings in the treetops. Males sing throughout the day, but I have rarely seen a bird remain in one of his singing posts for any length of time.

The female Tennessee is one of the quietest and most secretive of Wood Warblers. The lycopodium around the nest is typical of the kind of cover chosen for a nest site.

A Tennessee Warbler (left) and an American Redstart engage in a dispute over rights at a watering spot.

In ideal habitat, Tennessee Warblers sometimes nest in small, loose colonies. Damp forest openings are typical.

The female is a different story. She is one of the quietest, most secretive warblers that I encounter. I do not recall ever seeing a female Tennessee in a tree even when feeding. And if there is any courtship activity between the pair, other than the male's robust song, I have never witnessed it. Perhaps the flight song that David E. Baker watched in a black spruce bog in Chippewa County, Michigan, was a courtship display. The song was uttered sixty feet above the ground. "To my ears, the perch and flight songs are identical," Baker wrote.[1]

Favorite nesting haunts are cedar and larch swamps, bogs, thickets of young firs, edges of clearings or burns, and muskeg borders where grass and low bushes provide protection. The nest is on the ground, often in a mossy hummock and typically hidden from above by clumps of overhanging grass. Eggs of the Tennessee Warbler are very similar to those of the Nashville Warbler, but the Tennessee commonly lays four to seven eggs, often six, while the Nashville rarely lays more than five.

[1] Baker, David E., "Tennessee Warbler Nesting in Chippewa County, Michigan," *Jack-Pine Warbler,* vol. 57, no. 1 (1979), p. 25.

Female Tennessee Warblers lay larger clutches than most boreal warblers except the Cape May Warblers, which nest in trees.

Dr. Friedmann labels the Tennessee a very uncommon victim of the cowbird. Most of his few records are from Canada. Much of the range of this northern nester is at the northern edge of the Brown-headed Cowbird's breeding range or beyond. Of eight nests I have observed in Maine, none has been parasitized.

Early ornithologists often observed that the Tennessee Warbler is gregarious in a suitable nesting habitat. Two observers in Quebec told Arthur Cleveland Bent that they found sixteen nests of the Tennessee in a corner of a sphagnum bog, and, "there must have been about 100 pairs nesting in this ideal spot at the time."[2] George H. Harrison told me of colonies he observed on Bonaventure Island off the Gaspé Peninsula in June. My experience on Mount Desert Island, Maine, indicated only single nestings in widely scattered areas; however, on a field trip to Aroostook County, Maine, I found a colony of Tennessees in a small but ideal nesting spot.

[2] Bent, Arthur Cleveland, *Life Histories of North American Warblers* (Washington, D.C.: U.S. National Museum Bulletin no. 203, 1953), p. 79.

I photographed Tennessee Warblers nesting in Maine as late as August 2, when the young were four days old. Note fuzzy down typical of altricial young and pinfeathers on wings of the nestlings.

While most records of nests have been for June, Ralph Long has found the Tennessee Warbler uncommon on Mount Desert Island early in the season, but has observed annually a flurry of activity in July and even August. He has records of adults feeding nestlings on August 13 and 14. I photographed a nest on Mount Desert on August 2, when the young were four days old. In apparent contradiction, Robert Leberman at the Powdermill Nature Reserve in western Pennsylvania, south of where the bird is known to breed, has netted and banded thirty-nine adult female Tennessee Warblers in late July. All had brood patches and all were molting.

Dr. Kenneth C. Parkes has suggested to me that since it is uncommon to find such conditions in migrating birds, it is possible that these warblers came from some nearby, as yet undiscovered breeding area south of the presently known range. Despite nestings and singing males observed late in the season in the Tennessee Warbler's breeding area, it is considered unlikely that the species has two broods in a single year. Late nests may well be second or even third attempts after early nests have been destroyed.

These warblers have been accused of damaging grapes on their southward migration. During the nonbreeding season, Tennessee Warblers are largely nectar or fruit eaters. In their winter home in Panama they favor nectar from *Combretum fructicosum,* a small bush or a vine covering the crown of a tree. This plant is pollinated almost exclusively by the warbler. The face and throat of a bird feeding on combretum soon become red with the plant's pollen.

In 1832 when he discovered this warbler, Alexander Wilson happened to be on a collecting trip in Tennessee. As with the Nashville, which Wilson discovered earlier, the name is based only on where the bird was first seen. Neither bird nests in Tennessee.

Peregrina (per-reh-GRIN-ah) is from the Latin *per,* "through," and *ager,* "the land," and is translated as "across the country" or "wanderer." Coupled with *Vermivora,* it literally means "wandering worm-eater." Wilson thought it rare, possibly a wanderer far from its normal range, hence the name *peregrina.*

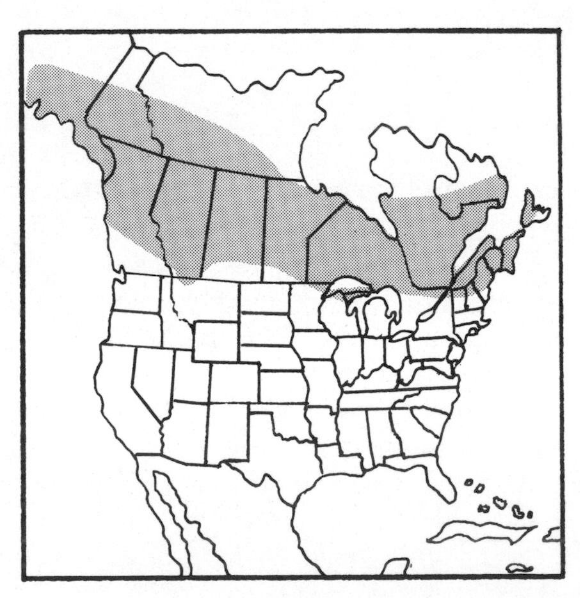

Breeding Range of Tennessee Warbler

5. *Orange-crowned Warbler*

Vermivora celata

PLATE 2

My most vivid memories of the Orange-crowned Warbler are of days replete with frustration spent in Temescal Canyon in the Santa Monica Mountains near Los Angeles. It was late April; males on their territories were singing their monotonous chippielike trill. Females were evident as they fed in low bushes along the trails. The time seemed right for nests, but not once in several days of watching and searching the leaf-covered hillsides did I see any evidence of nesting.

Throughout its vast breeding range, there are four recognized subspecies or races. The one I watched was the Lutescent Orange-crowned Warbler (*V. c. lutescens*), the most brightly colored of the group. I saw for the first time the well-concealed orange patch on a male's head when he ruffled his feathers. Females may or may not have this patch, but I saw none.

Celata (see-LAH-tah), the species name for the Orange-crowned Warbler, is Latin for "concealed" or "hidden," and refers to this orange that is hidden under the olive tips of the head feathers.

Attempts to identify the Orange-crowned Warbler in the eastern United States in autumn leave me just as frustrated and confused as they do most bird watchers. This tiny bird is drab and attracts little attention although it is probably much more common than we realize. Peterson calls it "the dingiest of all warblers."[1] To me, it is the most confusing of all the "confusing fall warblers."

[1] Peterson, Roger Tory, *A Field Guide to the Birds,* 2d ed. (Boston: Houghton Mifflin Company, 1964).

Female Orange-crowned Warblers may or may not have the well-concealed orange head patch.

The species was discovered and named by Thomas Say in May, 1823, along the Missouri River during spring migration. The bird he saw was an uncommon straggler, for Orange-crowns are unusual east of the Rockies in spring and are not at all abundant in fall.

Orange-crowned Warblers are wanderers and occasional winter residents in the east. Wintering birds are sometimes attracted to feeding stations by suet, peanut butter, and doughnuts. Herbert L. Stoddard attracted migrants to his Georgia feeding trays with pecan bits. His records indicate visitors from December to March. The birds winter regularly in the southern Atlantic and Gulf states, south to Baja California and Guatemala. I have occasionally seen single Orange-crowned Warblers in flocks with other warblers wintering in southern Florida, but they are not common.

The nesting habitats the species prefers include hillsides, brushy woodland clearings, chaparral, burned-over areas, overgrown pastures, and wooded edges of low deciduous growth. Nests typically have the rim flush with the ground and are rather large for the size of the bird. An exception to this is the nest of the Dusky Orange-crowned Warbler (*V. c. sordida*), a subspecies that breeds on the southern California mainland and on coastal islands, especially the Santa Barbara Islands. Females of this race build their nests in low bushes instead of on the ground.

In April 1923, a Dusky Orange-crowned Warbler built its nest in a decorative fern basket inside a lath house adjoining the A. P. Johnson home in urban San Diego. The nest contained three well-incubated eggs when discovered by Mrs. Johnson as she watered the fern. The

East of the Rocky Mountains, the Orange-crowned Warbler is an uncommon straggler in spring and not at all abundant in fall.

For nesting, Orange-crowned Warblers prefer hillsides, brushy woodland clearings, chaparral, burned-over areas, and overgrown pastures.

The nest and eggs of an Orange-crowned Warbler.

eggs hatched and the young left the nest thirteen or fourteen days later. The following year, a Dusky Orange-crowned built a nest in the same basket adjoining the first nest. The female was so fearless that she permitted herself to be lifted from the nest.

The male's song is somewhat like that of a Chipping Sparrow but higher in pitch and ending with two lower notes. W. E. Clyde Todd described it as "generally similar to those of the Nashville and the Tennessee Warblers and intermediate between the two.[2]

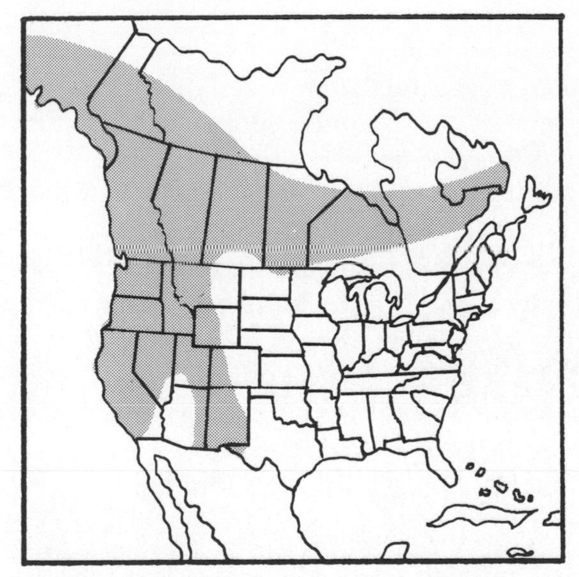

Breeding Range of Orange-crowned Warbler

[2] Todd, W. E. Clyde, *Birds of Western Pennsylvania* (Pittsburgh: University of Pittsburgh Press, 1940), p. 499.

6. *Nashville Warbler*

Vermivora ruficapilla

PLATE 3

It is impossible to describe the preferred nesting habitat of the Nashville Warbler in a simple, definitive sentence. After studying this species over much of its eastern breeding range, I am no longer surprised when I find a nest in an environment completely unlike others I have seen. Many nests I have studied have been on Mount Desert Island, Maine. Here they are commonly in sphagnum bogs, alder swales, and mossy forest edges. Yet, I'm not surprised when I find a nest in an open blueberry field, on the littered floor of a spruce forest, or along the edge of a forest path. At State College, Pennsylvania, I found the bird to be common in the dry understory of a pine forest known locally as "the pine barrens." In the Cheat Mountains of West Virginia, Nashville Warblers nest on the floor of overgrown forest openings at high elevations.

Nashville Warblers have a propensity for hiding their nests completely from view under clumps of grass, blueberry bushes, club moss, bunchberry, ferns, or similar plants. I have never found a nest without pushing cover aside. To see a nest with eggs without flushing the incubating female is virtually impossible, although it has been done by means of a down-on-the-knees search. In fact, I must admit that I have had to resort to this method even when I have flushed the female from her eggs. Literally crawling about the area from which the bird had taken flight has been necessary on several occasions before finding the completely hidden nest.

The four or five brown-spotted eggs are similar to those of Tennessee Warblers but slightly smaller. A normal clutch for the Tennessee is six; the Nashville rarely if ever lays more than five. At Nashville

Blue-winged Warbler pair at nest

PLATE 2

Singing male Bachman's Warbler (endangered species)

Colima Warbler

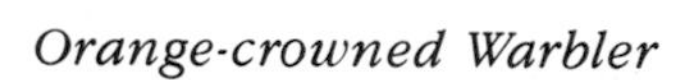

Orange-crowned Warbler

Nashville Warbler

Golden-winged Warblers, female, left; male, right

PLATE 4

Male Black-throated Blue Warbler

Female Black-throated Blue Warbler

Magnolia Warblers, male, left; female, right

Virginia's Warbler

Lucy's Warbler at nest cavity

Tennessee Warblers, male, top; female, bottom

Pair Chestnut-sided Warblers

Male Cape May Warbler

Northern Parulas, male, top; female, bottom

Male Yellow-rumped (Myrtle) Warbler

Female Yellow-rumped (Audubon's) Warbler

Pair Townsend's Warblers

The nest of a Nashville Warbler is hidden from view in this rank growth of ferns, cranberries and blueberries on Mount Desert Island, Maine.

nests I have observed, the female has assumed most responsibility for feeding young. She also seemed more concerned than her mate in removing fecal sacs. This is not true of all warblers I have studied. Among American Redstarts, for instance, males often did most of the feeding. In other cases parents assumed equal shares of the work.

The Nashville Warbler appears to be almost wholly insectivorous. Franklin Herbert Mosher watched a Nashville devour forty-two gypsy moth caterpillars in thirty minutes. Most caterpillars, both hairless and hairy, such as gypsy, browntail, and tent caterpillars, are fed to nestling Nashvilles.

Of fourteen Nashville nests I have found with eggs, the only parasitized nest contained four Nashville Warbler eggs and one Brown-headed Cowbird egg. Friedmann listed the Nashville as a decidedly uncommon victim of parasitism.

During spring migration after leaving their winter home in southern Texas, Mexico, and Guatemala, Nashville Warblers cross the Gulf

Nashville Warblers nest in a wide variety of situations. Here, a pair has chosen a bed of sphagnum moss with cover predominantly of blueberry. Notice the fine grass lining.

or travel inland and fan out to their breeding grounds in southern Canada and the northern border of the United States. During spring migration in 1808, Alexander Wilson first saw this species near Nashville, Tennessee, and named it for that city. Early ornithologists considered the bird rare. Audubon collected only a few in Louisiana and Kentucky.

Observers have compared the song of the Nashville to no less than seven other species: Tennessee, MacGillivray's, Yellow, and Chestnut-sided Warblers, Indigo and Lazuli Buntings, and Chipping Sparrows. I think it is similar to the song of the Tennessee Warbler. It is two-parted; in the first part a segment is repeated four to seven times; in the second, a segment is rendered three to five times in a descending trill. The second part is sung more rapidly than the first. Since the entire song lasts only about three seconds, it is difficult to count the number of times a segment is repeated in each part.

In their classic work, *The Birds of Arizona,* Allan Phillips, Joe Marshall, and Gale Monson propose that the Nashville, Virginia's, and Colima Warblers are too much alike to be individual species, and they present the group as a single species. Other ornithologists reject the "lumping" proposal. Allen K. Brush and Ned K. Johnson cite differences between the Nashville and Virginia's Warblers that they believe suffice to prevent intergradation should an expansion of range bring about contact of breeding populations.

The species name for the Nashville, *ruficapilla* (rue-fih-cap-ILL-as), is Latin for "rufous-haired," which alludes to the chestnut crown.

A male Nashville Warbler brings food to his mate, who will transfer it to the nestlings she is brooding. A downy body can be seen in front of her breast.

The crown patch is not very conspicuous in the male and is less so or lacking in the female.

In *Birds of the Connecticut Valley in Massachusetts* Samuel A. Eliot, Jr., acknowledges the misnaming of the Nashville, but his suggestion, while more apt, was hardly an improvement: ashy-coif (pale gray cap or hood). Ludlow Griscom, another scientist who decried the English name of the Nashville Warbler, went further and scoffed at the scientific name as well. He called it absurd and inaccurate. Translated, it means "rufous-haired worm-eater." Griscom noted that members of the genus *Vermivora* do not eat worms any more than any other warbler and that the species has very few rufous hairs.

In a letter to me, Dr. Kenneth C. Parkes, a member of the American Ornithologists' Union Check-list Committee, made this comment regarding the use of names: "Bird names are best considered as proper names even if not literally appropriate, just as most people named

Note the head of this brooding female Nashville Warbler. The chestnut crown patch is not very conspicuous in the male and is less so or lacking in the female.

Baker or Taylor do not follow these professions. They are too well established, in most instances, to be changed because of inappropriateness. Of the 100-plus changes in English names effected in the sixth edition of the American Ornithologists' Union *Check-list of North American Birds* (1983), none or almost none are changed on the sole grounds of inappropriateness."

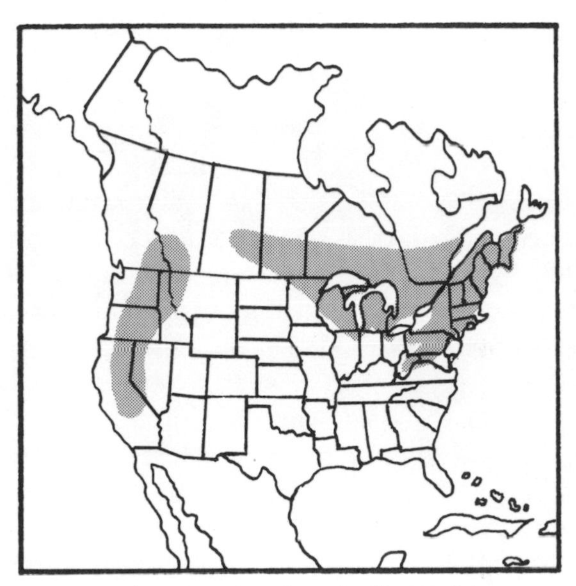

Breeding Range of Nashville Warbler

7. *Virginia's Warbler*

Vermivora virginiae

PLATE 5

Of the hundreds of Wood Warbler nests I have seen, few have been as difficult to find as those of Virginia's Warbler. The only time I can remember flushing any incubating Wood Warbler and being unable to locate its nest was in the Chiricahua Mountains, Arizona, where I found the first of three Virginia's nests. The bird left her nest in a mountainside of scree that was covered by layers of leaves. The footing was treacherous; a small landslide accompanied almost every step. A cautious and lengthy search disclosed no nest. I had about concluded that I had flushed a feeding bird. My companion, Dale Dalrymple, saved the day by finding the nest with four eggs buried deep in leaves, completely concealed from above. The entrance was a mere slit in the layer of leaves around it.

My second nest on the same mountainside was found later the same day, May 27, 1979, by my wife, Mada, and Blake Gardner. It too held four eggs. Both nests were on steep talus slopes above Barfoot Park at an elevation of about eight thousand feet.

The third was on a steep leaf-covered bank that rose abruptly from a small branch of Cave Creek in the Chiricahuas. The adults refused to go near it while we watched. The instinct to feed almost always overcomes the fear of intruders, so for warblers in general this behavior is unusual. When we finally located the nest, four fully feathered young fledged in one scrambling exit.

From its winter home in Mexico, Virginia's Warbler moves north in spring to the southwestern United States. It prefers chaparral in foothills and mountains, scrub oak canyons, and pinyon-juniper brushlands. In the foothills of the Rockies in Colorado, Virginia's is considered one of the most common Wood Warblers.

Nests I saw in the Chiricahuas are typical: on or in the ground, commonly embedded in dead leaves and often hidden by overhanging grass. Nests are built of strips of inner bark, grass stems, roots, and mosses. Both parents feed the young, and my observation is that they share responsibility equally. Many times as I watched, the two birds came to the nest simultaneously, both with food for the nestlings.

Possibly because the Virginia's altitudinal range—six thousand to eight thousand feet—is largely above that of the Brown-headed Cowbird, parasitism is rare.

Although the Virginia's song is two-parted like that of the Nashville Warbler, it is distinctive. The first part is a very rapid series of syllables, *tis, tis, tis, tis, tis;* the second series is shorter and slightly higher-pitched, *see, see, see.* Another interpretation of the song is *che-we-che-we-che-we-che-we, wit-a-wit-wit-wit.*

My experience in trying to photograph adults as they fed nestlings

Virginia's Warbler's nest is very difficult to find. It is generally buried beneath a layer of leaves with a mere slit for an opening.

Nesting habitat of Virginia's Warbler at Barfoot Park, eight thousand feet up in the Chiracahua Mountains, Arizona.

at a Barfoot Park nest is evidence of their extreme shyness. Although I was hidden in a blind fifty feet away operating by remote control a camera set a few feet from the nest, the parents refused to feed. I moved the camera back to ten feet from the nest, too far for photography, but I was confident that this would reassure the birds. It did not work. I finally removed the camera for several hours while the adults resumed feeding the five-day-old young. The next day, resuming my efforts, I placed the camera and tripod forty feet from the nest, where the birds accepted it. I moved it a few feet closer every hour or so. On the third day of patient waiting I had the camera close enough to the nest to take pictures. I remember no other bird that difficult.

I observed an unusual happening while waiting to photograph this pair of Virginia's Warblers. I caught sight of a House Wren entering their nest. A moment later it came out carrying a fecal sac and flew away. Later the wren returned and entered the nest quietly and stealthily. This time it came out with nothing. At no time did the warblers appear to resent the wren at the nest or in their territory.

Virginia's Warbler was discovered in 1858 in New Mexico by Dr. William W. Anderson, an assistant surgeon in the U.S. Army at Camp Burgwyn. It was named for Anderson's wife by Professor Spencer F. Baird, who described the species in 1860.

Some ornithologists believe Virginia's, Nashville, and Colima Warblers belong to a single species. This debate is discussed in chapter 6.

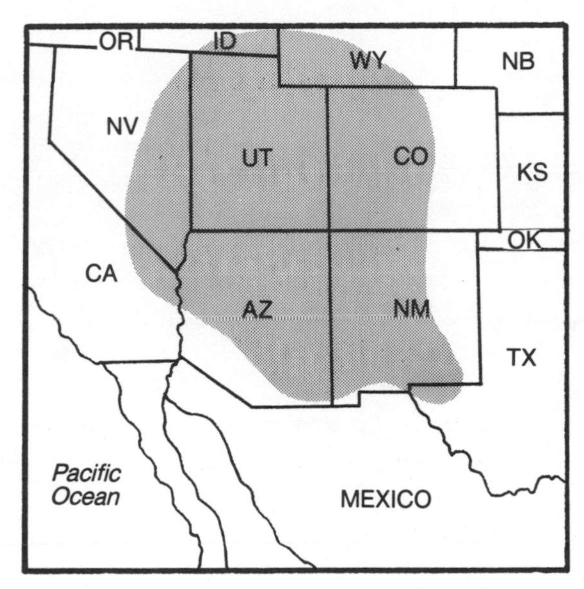

Breeding Range of Virginia's Warbler

8. Colima Warbler

Vermivora crissalis

PLATE 2

If you don't fancy riding horseback or hiking up a rough mountain trail to Boot Spring, high in the Chisos Mountains of Big Bend National Park, where the Colima Warblers come from Mexico to Texas to nest every spring, take heart; you may add this species to your life list without such a struggle. You won't see a nest, for the birds prefer to breed in the oak-clad mountains at an altitude of over six thousand feet; but Roland H. Wauer, of the National Park Service, declares that bird watchers can usually find most of the same birds at Laguna Meadow and its canyons in the same park. Wauer says that another area that can be good in spring, summer, and fall is the Lost Mine Trail. "This trail starts at 5,800 feet elevation and a two-mile walk offers some high-country exposure for the minimum of time," he adds.[1]

Before May 7, 1932, a list of Wood Warblers known to nest in the United States would not have included the Colima. Over fifty years after its discovery, its known breeding range in our country is still confined to this very limited area in the Chisos Mountains. Its principal breeding range is in the mountains of northeastern Mexico. Although it was described in 1889 from a specimen collected by W. B. Richardson in the Sierra Nevada de Colima, Mexico, less than a dozen specimens were known when Frederick M. Gaige collected the first for the United States in 1928.

In 1932 in the Chisos Mountains Josselyn Van Tyne watched a

[1]Wauer, Roland H., *Birds of Big Bend National Park and Vicinity* (Austin: University of Texas Press, 1973), p. 32.

Breeding range of the Colima Warbler in the United States is limited to the Chisos Mountains in Big Bend National Park, Texas.

female carry building material to a nest lodged between two rocks and deeply imbedded in dead oak leaves on the bank of a dry stream. A dense ground cover of vines and other herbaceous plants arched completely over it. The description of this nest and its location, and the account of the four eggs the female subsequently laid, could serve as a nesting report for the closely related, very similar Virginia's Warbler. A year after Van Tyne's discovery, George M. Sutton discovered the second nest known to science.

Sutton described the Colima's song as a "Chipping Sparrow-like trill." Van Tyne's description was similar. "The common song of the Colima Warbler is a simple trill, much like that of the Chipping Sparrow but rather shorter and more musical and ending in two lower notes."[2]

[2] Van Tyne, Josselyn, *The Discovery of the Nest of the Colima Warbler* (*Vermivora crissalis*) (Ann Arbor: University of Michigan Miscellaneous Publications no. 33, 1936), p. 7.

Both parents help care for the young. Notice how this nest is hidden from above by an overhanging clump of grass. The Colima's nest and eggs are very similar to those of Virginia's Warbler.

Emmet Blake discovered and photographed this secluded nest of a Colima Warbler.

More recently, annual counts of Colima Warblers were taken in the Chisos during the second week of May. All locations where suitable habitat exists were searched. The results: 92 in 1967, 130 in 1968, 166 in 1969, and 118 in 1970. In all instances, the birds were associated with oak-pinyon-juniper or oak-maple-Arizona cypress environments.

The species name, *crissalis* (kris-ALE-iss), refers to the bright color of the crissum, or under-tail coverts. *Colima* (Coe-LEE-mah) is for the state in southwest Mexico where the type specimen was collected.

Breeding Range of Colima Warbler

9. Lucy's Warbler

Vermivora luciae

PLATE 5

My notes on the behavior of nesting Lucy's Warblers were made during two seasons in Arizona. In 1975, in Florida Wash at the base of the Santa Rita Mountains in Pima County, I found four nests, all within six feet of the ground, and all in dead branches or knotholes in trunks of mesquites. In 1979, I investigated areas in Cochise County. One was a sycamore grove along lower Cave Creek where I found nests twelve, twenty, thirty, and forty feet above ground in sycamores. Another area was an arid desert wash called Round Valley. Here I found four more nests, mostly in cracks and cavities between three and eleven feet above ground in desert willows.

Four cavity nesting sites of Lucy's Warbler in southeastern Arizona.

M. French Gilman, reporting on twenty-three nests of Lucy's Warblers found along the Gila River bottom in Arizona, lists four types of nesting sites in this order of frequency: in natural cavities, under loose bark, in woodpecker holes, and in deserted Verdin nests. Occasionally, Lucy's Warbler will build in a hole eroded in the bank of a stream.

The nest built in these cavities is a small compact cup of fine grass, twigs, dry leaves, and plant fibers, some lined with horse or cow hair, some with feathers. The nest material is sometimes visible from outside the nesting cranny.

At a nest high in a sycamore at Cave Creek Ranch, I watched both birds carry nesting material into a natural cavity. They worked rapidly and industriously. Several times both arrived at the same time and they entered together. They were so similar that I was never able to determine if the male incubated or brooded. Other observers claim those chores fall entirely to the female.

Its nesting habits make Lucy's unique among paruline warblers. It is the only one that nests on the desert, and it is the only cavity-nesting warbler in the western United States. (An eastern species, the Prothonotary Warbler, also nests in cavities.)

I found Lucy's a very shy bird, difficult to flush from its nest. When

Lucy's Warbler is one of two Wood Warbler species that build their nests within cavities.

Lucy's is the only Wood Warbler that nests on the desert. Florida Wash at the base of the Santa Rita Mountains in Arizona is a typical habitat.

dislodged, the bird leaves quickly without a fuss. While photographing a pair feeding young, I noticed that what I took to be the female was more intimidated by the camera than her mate. She regularly hesitated before bringing food to the young. She hopped from branch to branch, obviously working up courage to approach the nest. Very often the male would physically nudge her in an apparent attempt to force her to feed. The female usually responded to the male's nudge. It happened so often that it became a bit of characteristic behavior.

Nest sanitation in this species has puzzled me. At several nests where Lucy's Warblers were feeding young, I have found a small pile of excrement at the front rim of the nest. Since the adults, like other paruline warblers, remove fecal sacs deposited by nestlings, and since this excrement was not in the form of fecal sacs, I wonder if it was deposited by the female as she incubated or brooded. At any rate, it is a condition I have observed at no other warblers' nests.

Cavity nesting does not seem to be a safeguard against the Brown-headed Cowbird. Friedmann lists quite a few cases where a cowbird has been able to squeeze into a cavity. W. L. Dawson cited the Gila Woodpecker as an enemy. He wrote: "Accustomed as he is to poking and prying, he seems to take a fiendish delight in discovering and devouring as many Lucy's Warbler eggs as possible."[1]

Lucy's Warbler winters in western Mexico south to Jalisco and Guerrero, and is an abundant summer resident in the cottonwood-mesquite habitat of the Lower Sonoran Zone and the sycamore-live oak association of the Upper Sonoran Zone. In the sycamore groves

[1] Cited in Bent, ed., *Wood Warblers,* p. 134.

Setup for photographing Lucy's Warbler at its nest site in a cavity in a desert willow, Round Valley, Cochise County, Arizona.

that border Cave Creek in the lower elevations of the Chiricahua Mountains and along the steep banks of that stream where ravine slopes approach the water's edge, I found Lucy's Warblers and Virginia's Warblers in proximity. Occasionally I was within hearing distance of singing males of both, and the songs were so similar that I could not identify the singer without seeing the bird. Lucy's song is two-parted and on two pitches. It is described as *weeta weeta weeta che che che che che che.*

Lucy's Warbler was discovered during the first year of the Civil War, March 25, 1861, by Dr. J. G. Cooper near Fort Mohave, Arizona. In describing the species, Cooper named it for Lucy Hunt Baird, the thirteen-year-old daughter of Professor Spencer F. Baird, former secretary of the Smithsonian Institution. The first known nest was found by Major Charles E. Bendire at Tucson, Arizona, May 19, 1872.

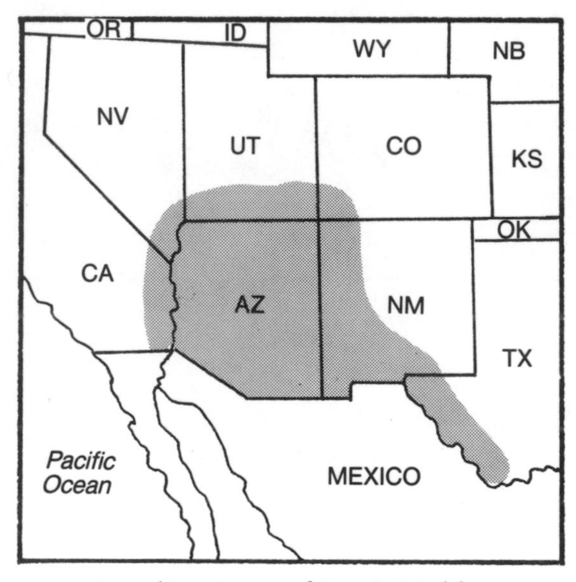

Breeding Range of Lucy's Warbler

10. Northern Parula

Parula americana

PLATE 7

We who have visited the Northern Parula in various parts of its extensive breeding range in the eastern United States and Canada are impressed with the adaptability of this tiny bird, one of the smallest of Wood Warblers. Whether nesting in the North where old man's beard (*Usnea* spp.), a tree lichen, is available, or in the South where it uses Spanish moss (*Tillandsia usneoides*), an epiphyte, or in between where neither is available, the Northern Parula survives.

While searching for Bachman's Warbler in the I'On Swamp in late March, I found Northern Parulas building nests in Spanish moss that hung like tattered gray curtains from ancient trees. Of the seven nests I located, five were in cypress and two in black gum. Heavily forested bottomland where there is much Spanish moss is a typical habitat. Either conifers or broadleaved trees are acceptable.

A typical nest in Maine is built into a clump of *Usnea* hanging from a tree in a rather open site. Seventy-one nests observed by Allan Cruickshank in Lincoln County, Maine, were all in *Usnea.* Sometimes the female adds more strands of the lichen and perhaps a little fine grass to the lining, but nearly always the nest is wholly of lichen. The completed structure is a gray purse or sack with an opening at or near the top. My wife is particularly adept at finding the Northern Parula's nest even in a tree loaded with old man's beard. She looks for the beard with the "tennis ball" in the bottom.

The adaptability of this species is most evident in areas where there are no lichens or moss. A nest has been found in a clump of drift grass caught in a maple branch overhanging a river; one composed entirely of leaf skeletons was found in a hemlock; another was of

A tree laden with Spanish moss is typical habitat for the Northern Parula in the South.

In New England, where *Usnea* (old man's beard) is available, Northern Parula build their nests inside hanging "beards."

Where Spanish moss is available, Northern Parulas invariably build their nests within this epiphyte. Note the fine, soft material that has been added.

skeletonized leaves and pine needles. In Georgia's Piedmont region, the Northern Parula often constructs its nest in a wooded ravine of hemlock where a light-green mosslike lichen is available as a substitute for *Tillandsia.* I found an unusual nest in the mountains of West Virginia where an ingenious bird had built inside a piece of burlap snagged in a hemlock branch.

The height of the nesting site is apparently not important. Eighteen nests I have found in Maine have ranged from four to forty feet above ground, averaging twenty-five feet. The seventy-one nests found by Cruickshank ranged from five to fifty-four feet above ground.

The same nesting site is often occupied in successive seasons, eggs being laid in the same nest or in another nearby. Ralph Long reports a nest, fifteen feet above a trail on Mount Desert Island, Maine, which had been used for four successive years.

Henry Mousley gathered some interesting statistics in long hours of patient observations in Quebec. In three and a half hours a female

Where neither Spanish moss nor *Usnea* is available, Northern Parulas often nest in strange places. Here is a nest in a piece of burlap lodged in a tree.

brought sixty-two loads of material to a nest she was building. In another two and a half hours she brought forty-three loads. Although the male never carried anything to the nest, he accompanied his mate on 111 of 221 journeys. These trips were made at the average rate of one every 4.6 minutes. At another nest that Mousley studied for twenty-four hours between May 22 and 31, the female carried 206 loads of nesting material. The male carried nothing to the nest, but he sang 549 times during that period.

Northern Parulas in other areas must be more successful in their nesting attempts than those that I have studied on Mount Desert Island. From what I have noted there, my conclusion is that a Northern Parula pair is lucky to bring off a brood of four. The cowbird is not a problem in that area, and the Northern Parula is rarely parasitized; but an inordinate number of nests that I have found have been destroyed. In most cases, the nests have been badly torn, which leads me to believe that the culprit is the red squirrel, an abundant mammal in Maine.

In fall, Parula families join the hordes of other warblers heading toward their winter haunts. They now visit deciduous woods in search of food. Countless thousands pass through Florida en route to the West Indies while others stream westward toward Mexico and Central America. The southward journey reaches its peak in September and continues into October.

The name *Parula* means "little titmouse." It comes from the Latin *parus,* "small," and is the generic name of the chickadees and titmice. The name was applied to the Parula because it searches under foliage for insects like a chickadee or titmouse. Audubon and Wilson called it the Blue Yellow-backed Warbler; Mark Catesby, the Finch Creeper. Linnaeus named the species in 1758 from a painting by Mark Catesby.

The Northern Parula's song is so diagnostic that one does not need to see the bird to know it is near. No other warbler song has such an

A Northern Parula's nest in the dangling twigs of a hemlock tree is a more open cup, different from the hidden nests found in the North and South. This nest was in West Virginia.

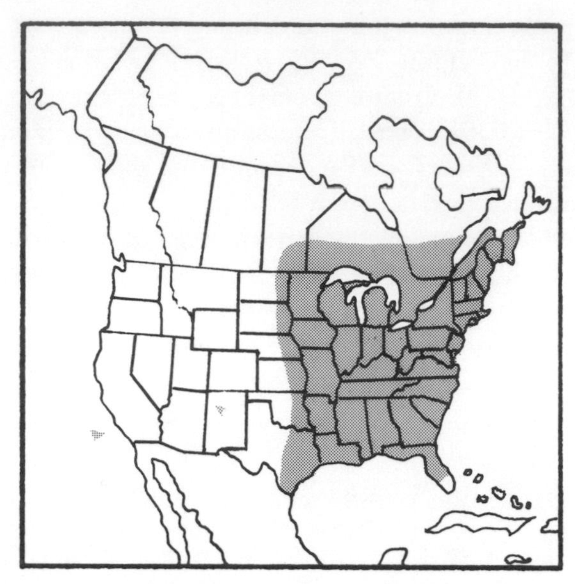

Breeding Range of Northern Parula

explosive ending. The first part is a prolonged ascending trill. The ending is a very abrupt lashlike note. There is a tendency to increase in volume toward the end. C. J. Maynard offered this description of the song: *swee swee swee swee swee-zer.* Alexander Sprunt, Jr., called it "a buzzy ascending trill ending in a sharp emphatic *zip.*"[1] Parkes offers a unique but very descriptive interpretation: "like winding a watch too tight until it breaks with a snap."[2]

From time to time I hear someone pronounce the name Pa-ROO-la. Both Webster's and Random House dictionaries and *The Biologist's Handbook of Pronunciation* by Edmund C. Jaeger, D. Sc., give the pronunciation as PAR-a-la or PAR-ya-la. Take your choice.

A male and a female Wood Warbler, discovered by Karl W. Haller and Lloyd Poland on May 30, 1939, south of Martinsburg, West Virginia, were described as a new species, Sutton's Warbler (*Dendroica potomac*). They are now believed to have been hybrids of the Northern Parula and the Yellow-throated Warbler.

[1] Sprunt, Alexander, Jr., and E. Burnham Chamberlain, *South Carolina Bird Life,* rev. ed. (Columbia: University of South Carolina Press, 1970), p. 98.

[2] Personal correspondence.

Young Northern Parulas are similar to but duller than the adult female. There is less yellow below and no sepia on the breast.

11. *Tropical Parula*

Parula pitiayumi

NO PLATE

As a nesting bird, the Tropical Parula reaches the United States only in or near the lower Rio Grande Valley of Texas. Even there it is a rarity. A pair nested in Bentsen-Rio Grande Valley State Park in 1981, the first known nesting in the Rio Grande Delta since 1966. At this time the only spot known to harbor annually a few nesting pairs is a roadside rest area on U.S. Route 77, about three miles south of Sarita, Kenedy County. However, a Tropical Parula sighted in Garner State Park in Uvalde County on April 19, 1980, may signify the beginning of a range extension. This is by far the deepest the bird has so far penetrated into the United States.

The Tropical Parula's future as a breeding bird in this area must be considered tenuous. Contributing to its present scarcity are such factors as cowbird parasitization, the effect of pesticides, and continuing disappearance of Spanish moss, the birds' favorite nesting site.

There is little to distinguish the Tropical Parula from its eastern counterpart, the Northern Parula. The male lacks the dark breast band and conspicuous white eyelids of the Northern, and has the added feature of a black area on the lores and cheeks. The song, nest, eggs, and habitat are similar to those of this closely related species. Like the Northern Parulas that nest in the Southeast, Tropical Parulas use Spanish moss to conceal their nests. Occasionally a nest has been found buried in an orchid or a dangling cactus.

Most of the birds that reach southern Texas in the nesting season (April through July) retire to Mexico, Central America, and South America to spend the winter months with other Tropical Parulas that

The Tropical Parula reaches the United States as a nesting bird only in or near the Rio Grande Valley in Texas. Spanish moss is available here as a nesting site.

are permanent residents in tropical woodlands and river bottoms. It is believed that many of these resident birds remain mated throughout the year.

The species name, *pitiayumi* (pit-ih-ah-you-MIH) derives from a Paraguayan Indian name for the bird. Since its discovery, the species has been labeled Sennett's Warbler, Olive-backed Warbler, Pitiayumi Warbler, Tropical Parula Warbler, and, now, Tropical Parula.

George B. Sennett was the first to record the species as an inhabitant of the United States when he collected a specimen at Hidalgo,

Texas, on April 20, 1877. Three months later the first nest was found near Brownsville, Texas, by Dr. James C. Merrill.

The Tropical Parula is one of three Wood Warblers whose breeding range in the United States is limited entirely to Texas. The others are the Colima Warbler and Golden-cheeked Warbler.

Breeding Range of Tropical Parula

12. *Yellow Warbler*

Dendroica petechia

PLATE 9

When I was a little boy growing up in a small town in western Pennsylvania, I am sure that I was well acquainted with the Yellow Warbler long before I knew it was a warbler. Like the rest of my companions, I usually called it a wild canary, sometimes just the yellowbird. One June morning, while catching butterflies in Mrs. Stofield's flower garden on the bank of the Allegheny River, I found the cottony nest of a yellowbird in a thorny rosebush. When I approached, the female flew away, exposing her four spotted eggs.

Even the most casual bird watcher is likely to know this species, one of the most widely distributed of all North American Wood Warblers. At all seasons this warbler is the yellowest of its family. Many people know the bird because of its broad breeding range, but more importantly because it is a conspicuous garden bird and is not as shy as most warblers. To see it requires little effort, for its brilliant yellow plumage demands attention when it flashes across a dooryard or garden. This is especially true in the East, where suitable habitat is widespread. In the West, Yellow Warblers are found principally in the river bottoms.

Males arrive on nesting areas ahead of females and begin immediately to defend territories. Increase in singing is the rule when prospective mates arrive. Courtship time is short and nest building begins soon. The female works alone and needs three to five days to complete a nest.

While many pairs nest in shrubs and bushes in city yards and parks, others seek more secluded environments along waterways, edges of swamps, marshes, brushy bottomlands, small trees, orchards, and roadside thickets. It has been my observation that multiflora rose offers ideal nesting sites. In such a desirable area, Yellow Warblers may be found in colonies and within territories that are remarkably small.

I was surprised to discover that, in a dense colony of Yellow Warblers, females stole material from other nests. In one case, removal of material was from a female's first nest, which she had recently deserted because of parasitization. In another instance, a female removed all material from an empty nest and added it to the one she was constructing.

Nests are placed in upright forks or crotches of shrubs, trees, or briers. The average height above ground is from three to six feet; some have been reported between two and twelve feet, and a few (in prairie cottonwoods, for instance) as high as forty to sixty feet.

The male occasionally feeds the female during the eleven or twelve days she incubates. Yellow Warblers are particularly fond of caterpillars which when plentiful form about two-thirds of the birds' diet. Fondness for larvae of the gypsy moth, tent caterpillars, cankerworms, and other measuring worms makes the Yellow Warbler very beneficial.

One of the first nests I found as a boy was the cottony nest of the "yellowbird" (Yellow Warbler).

A field of multiflora rose is ideal habitat for nesting Yellow Warblers. Photography from this blind was successful. Territories were relatively small and nests close together.

During 29 hours of watching in a ten-day period, Henry Mousley saw the male Yellow Warbler feed the nestlings 112 times and the female feed them 277 times, a rate of once every 4.5 minutes. At another nest watched by H. C. Bigglestone, the parents fed the young on 2,373 occasions during 144 hours of observation, an average of once every 3.7 minutes. The female fed almost twice as often as the male.

The Yellow Warbler is undoubtedly one of the most frequent victims of the Brown-headed Cowbird. Friedmann stopped counting when his files listed over nine hundred instances. On some occasions it may be a tolerant host, but at other times the Yellow Warbler eliminates the foreign eggs by building a new lining or nest floor, leaving them buried in the structure. There are numerous records of two-, three-, four-, and five-storied nests of Yellow Warblers, each of the lower stories containing cowbird eggs and some eggs of the warbler

The Yellow Warbler is one of the most frequent victims of the Brown-headed Cowbird. This female ended her nesting cycle feeding a lone cowbird and none of her own young.

as well. A record was reported by Andrew J. Berger, who found a six-storied nest with a total of eleven cowbird eggs buried in the various layers.

The male Yellow Warbler does not confine himself to a single song. Lynds Jones offers four interpretations: *sweet sweet sweet sweet sweeter sweeter* or *sweet sweet sweet sweetie* or *wee-che wee-shee wee-i-u* or *wee-chee chee chee chur-wee.* My wife who has a more musical ear than mine, suggests *sweet-sweet-sweet-chit-tit-tit-teweet.* I have difficulty differentiating between songs of the Yellow and Chestnut-sided Warblers.

Sometimes a Yellow Warbler avoids incubating the cowbird's eggs by building a new lining or nest floor, leaving them buried in the structure.

As soon as this youngster is able to take care of itself, the Yellow Warbler-family will begin its journey south to spend the winter in the tropics.

Mousley reported that during twenty-nine hours of observation while there were young in the nest, a male Yellow Warbler sang 27 times at the nest after feeding the young, 35 very near it, and 1,738 times away from it, making a total of 1,800 times, or a rate of more than once every minute. At times the song was repeated at the rate of six times a minute.

The Yellow probably spends as little time in the United States as any Wood Warbler. Time on the nesting territory may be as little as three months (May through July). They leave soon after the young are able to care for themselves. Although populations in the United States and Canada are migratory, there are related sedentary groups of Yellow Warblers that inhabit tropical areas from Mexico to the West Indies south to coastal Venezuela, Peru, and the Galapagos.

The Latin name of the Yellow Warbler refers to its favorite habitat. *Dendroica* (den-DROY-kah) is coined from the Greek *dendron,* "tree," and *oikeo,* "dwell"—a "tree-dweller." *Petechia* (peh-TEE-chi-ah) is Latin for "red spots on the skin," alluding to the chestnut streaks on the sides and breast of the male. This is another of the warblers first recorded by Catesby and Edwards, and then named by Linnaeus.

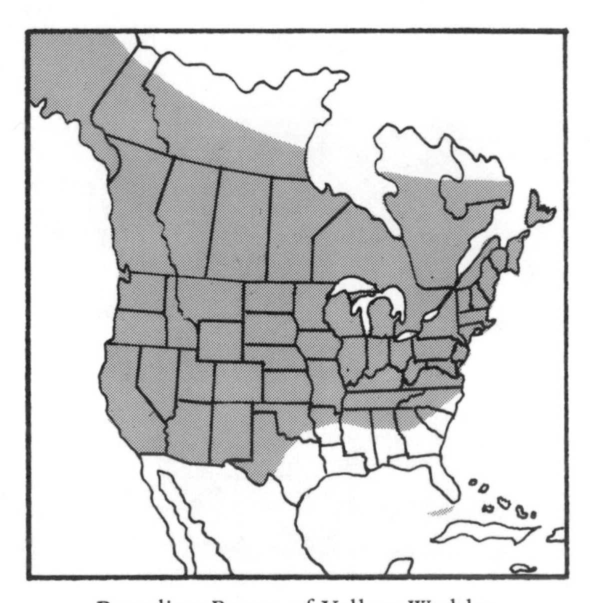

Breeding Range of Yellow Warbler

13. *Chestnut-sided Warbler*

Dendroica pensylvanica

PLATE 6

Of several thousand photographs that I have taken of wild birds, probably the most spectacular was made at the nest of a Chestnut-sided Warbler. It was taken at 4:34 A.M., May 25, 1946, in Butler County, Pennsylvania, when I caught a female Brown-headed Cowbird removing one of two warbler eggs from the warbler's nest. The picture was a bit of luck. I was not expecting to take an egg-removal picture. I had planned to catch the cowbird laying an egg in the warbler's nest. It had been my experience that the cowbird lays its egg in the host's nest at dawn. Usually, though not always, a cowbird selects a nest that already holds two eggs (possibly more). To leave one egg in the nest makes the theft less obvious than to leave an empty nest. Removal of one of the host bird's eggs commonly takes place sometime during the day before the cowbird lays its own egg, or occasionally may happen later in the day after she has deposited her egg in the nest. But not this time. The cowbird flew away with the egg lodged in her mandibles. Shortly thereafter, the female Chestnut-sided Warbler arrived at her nest, settled down inside, and laid her third egg. But her nest now contained only two. The perfect ending to this story would be to tell you that the Brown-headed Cowbird returned the next morning and laid its own egg, replacing the warbler egg it had stolen. I was there, ready for the historic event. It didn't happen. I never saw the cowbird again. The warbler laid her fourth egg and incubated and hatched three eggs—all her own.

The Chestnut-sided Warbler was practically unknown to early American ornithologists. Audubon saw only one bird; Wilson saw very few; and Thomas Nuttall, an English naturalist who came to the United

One of the most unusual and spectacular pictures ever taken by the author: a female Brown-headed Cowbird removing an egg of a Chestnut-sided Warbler to make room for a cowbird egg. This is the only such picture I know of.

Unaware of the loss of one of her eggs, the female Chestnut-sided Warbler returns to the nest vandalized by the cowbird.

States in 1808, considered it rare. That it has flourished and increased with the spread of civilization is apparent.

We are prone to view the coming of civilization to the North American continent as an adverse factor in the lives of wild birds, but such is not always the case. Chimney Swifts have found chimneys more abundant than hollow trees; House Wrens have benefited greatly from the thousands of bird boxes erected for their benefit; Barn Swallows not only have taken advantage of barns but have multiplied by the millions because of highway bridges that offer safe nesting sites; and Chestnut-sided Warblers have expanded their range and their numbers because of the increase of second-growth habitat created by man's clearing of the forests.

During the late nineteenth and early twentieth centuries, land that earlier had been cleared for farming but proved unsuitable was abandoned. The Chestnut-sided Warbler benefited greatly from an increase in the number of neglected and overgrown fields and pastures, where it found a very satisfactory breeding environment. Indeed, the

Thickets, briers, bushes, and brambles are optimum nesting habitat for the Chestnut-sided Warbler. Nests have been found in this edge between road and forest; the warbler disappears from areas where trees become dominant.

species is so closely allied to this habitat that it disappears from an area if trees become dominant and form a shady canopy. Today, in some areas, many old farmlands are overgrown to woodlands. As a result, some ornithologists believe that the Chestnut-sided, although still a common eastern warbler, may be less abundant than it was, say, forty years ago.

From a winter home in the lowlands of Panama and Central America, the Chestnut-sided Warbler migrates in early spring through eastern Mexico and the Gulf states from northwestern Florida to eastern Texas. As they move north through the Appalachians and the Mississippi Valley the birds spread out to occupy their breeding range. The

The Chestnut-sided Warbler has expanded its range and its numbers because of the increase of second-growth vegetation created when forests were cleared. This nest is in wild blackberry.

In fall plumage, the Chestnut sided Warbler is among those referred to as "confusing fall warblers"—difficult to identify.

Rockies seem to be a barrier to further extension westward, but a small colony of Chestnut-sided Warblers has been known from the eastern Colorado foothills since 1968.

The nest site naturally will vary with the plants available. For example, as reported by James Tate, Jr., seventeen nests found in the Douglas Lake region of Michigan were situated as follows: six in meadowsweet; three in raspberry; two in red osier; and one in each of six other plants. Of twenty-one nests of this species that I have found in Maine and Pennsylvania, the majority were in blackberry, spirea, or alder.

The nest a female Chestnut-sided Warbler builds alone in about five days is a loosely woven, thin-walled cup of inner bark strips, shredded weed stems, grasses, rootlets, and plant down. It is often attached to surrounding twigs with spider silk or wool. I have seen a few nests that were compactly built, but most are rather flimsy and loosely constructed. In some, the walls have been so thin that daylight showed through in places.

At one nest in Maine where I photographed adults feeding young, the male was much bothered by my camera. Although I photographed him early on, he eventually figured a way to avoid coming to the nest and thus exposing himself. With his mouth full of food, he lit several feet away where he waited for the female. When she came to him, he transferred the food to her mouth and flew away. His mate then took his offerings to the nestlings.

Although it is generally recognized that the female is similar to the male but duller in color, it has also been pointed out that it is sometimes hard to distinguish between sexes on appearance alone (the female, of course, does not sing.) This fact was brought home to me while I was taking pictures at the nest mentioned above. The two were so similar that the only way I could tell them apart was by the slightly brighter yellow cap on the male. Contrary to what one might expect, the chestnut on the sides of the female actually was brighter and more extensive than the color on her mate.

If the male Chestnut-sided Warbler had only one song in his repertoire, it would be quite easy to recognize it. The territorial or spring song has been interpreted as *I wish to see Miss Beecher* or *dis-dis-dis-dismiss you* or *very-very pleased to meetcha* or *sweet sweet sweet I'll*

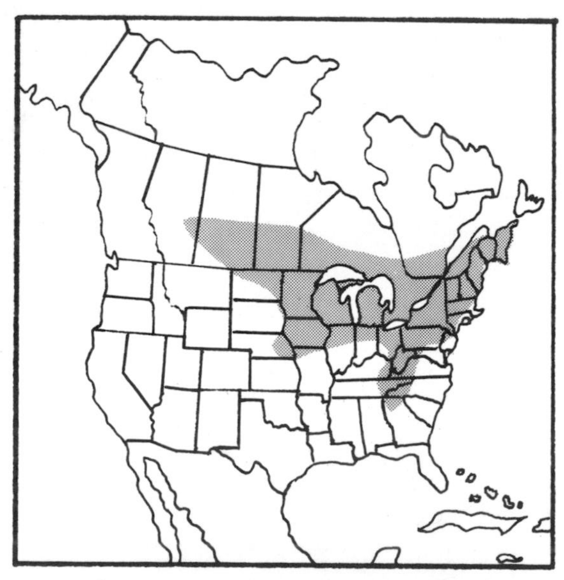

Breeding Range of Chestnut-sided Warbler

switch you. Emphasis is always on the ending, specifically the penultimate syllable. The notes have a similar tone quality to those of the Yellow Warbler. A second song, often referred to as the nesting song, is quite variable and less distinctive. It is a longer run of slurred two-note phrases that are generally lower in pitch.

Editors, copy editors, proofreaders, operators of word processors, and printers must be cautioned not to change the spelling of the scientific name, *Dendroica pensylvanica.* The uninformed decide that the writer has made a typographical error, that the name should be *pennsylvanica.* Not so! The first specimen was taken in Philadelphia, and in assigning its name, Linnaeus wrote *pensylvanica,* an obvious misspelling, meaning "of Pennsylvania." And it cannot be changed! The International Code of Zoological Nomenclature provides that the spelling of a species name must be preserved as given by the author even when the spelling is wrong or the word does not conform otherwise to standard usage.

14. *Magnolia Warbler*

Dendroica magnolia

PLATE 4

"The Magnolia Warbler is to my mind the most strikingly beautiful warbler that makes its home in New England. The Blackburnian with its orange front may be preferred by many, but that bright front is its chief glory, while the Magnolia Warbler's beauties are distributed to all parts of its graceful form."[1]

So wrote Edward Howe Forbush, famous New England ornithologist. You and I may or may not agree, but it is certain that this handsome black and yellow warbler will be high on anyone's list of beautiful birds. The female suffers in comparison to her mate only in having uniformly duller plumage.

Before I knew the Magnolia Warbler, I was well acquainted with the Hooded Warbler in my Pennsylvania woods. When I first heard a Magnolia singing in Maine, I was amazed that a "Hooded Warbler" should be singing so far north of its normal range. To me, one of the Magnolia's common songs can be interpreted as *ta-weet ta-weet ta-weeteo.* Roger Tory Peterson records the song as a loud whistled *weeta-wee-tee-o.* My wife has a unique interpretation. To her, the bird pronounces the name of a small Georgia town, *lewd-a-wee-chee* (Ludowici). To Kenneth C. Parkes, the best interpretation is *WEST vir-GIN-ia.* William Brewster, founder and president of the American Ornithologists' Union, offers his impression of the Magnolia's song in these words: *she knew she was right; yes, she knew she was right.* A shorter version by Brewster: *pretty, pretty Rachel.*

[1] Forbush, Edward Howe, and John Bichard May, *Natural History of the Birds of Eastern and Central North America* (Boston: Houghton Mifflin Company, 1939), p. 422.

Coniferous forests are home to Magnolia Warblers during the nesting season. A stand of interlaced young spruces is a favorite nesting habitat.

This warbler comes back to us in spring from its winter home in Mexico and Central America south to Panama. A few occasionally winter along the Gulf Coast.

One of the greatest tragedies to befall migrating birds occurred May 7, 1951, when a storm over the Gulf of Mexico accompanied by cold winds caused the death of thousands of birds on Padre Island, Texas. On May 8, a group of ornithologists gathered 2,421 of the thousands of dead birds and identified them. In all, thirty-nine species were represented; nineteen species were warblers and made up 95.35 percent of the total number of birds. Of those, 1,109 were Magnolia Warblers. Second in number were Common Yellowthroats.

Spruces, hemlocks, pines, and balsam firs of the Coniferous Forest biome are home to Magnolia Warblers during the nesting season. Usually they hide their nests well in the bushy tops or horizontal branches of small conifers. Of over fifty nests found in Warren County, Pennsylvania, by R. B. Simpson, all but one were in hemlock. One nest was at the exceptional height of thirty-five feet; another was only a foot from the ground in hemlock brush. Of thirty-three nests studied by Allan Cruickshank in Lincoln County, Maine, eighteen were less than five feet above ground, thirteen were between five and ten feet, and only two were over ten feet. The highest was fourteen feet. All were in red or white spruce or in balsam fir. Although the great majority of recorded nests have been in conifers, I have accumulated records of six nests that were built in spirea bushes.

The female Magnolia Warbler pays more attention to hiding her nest well than to building a sturdy structure. Generally a nest is so loosely laid among the horizontal twigs of the tree that it can be lifted intact from its position. It is made of fine grasses, fine twigs, and forb stalks. The lining invariably contains coal black material. In addition, nests I have found in Maine commonly have held red hair-cap moss stems in the linings. Some linings were startlingly red from the abundance of this moss.

Eliot Porter offers a tip to persons searching for the nest: "The Magnolia Warbler utters a peculiar squeaky alarm note when disturbed at its nest; since it is not usually heard at any other time, it

In over two hundred nests of the Magnolia Warbler found in northeastern Pennsylvania, the typical clutch was four eggs. The black rootlets used in the lining of this nest are also typical.

The female Magnolia Warbler pays more attention to hiding her nest well than to building a sturdy structure. It is loosely laid among horizontal twigs.

Of thirty-one Magnolia Warbler nests examined by the author in Maine and Pennsylvania, six were parasitized by cowbirds. One cowbird and one warbler survived in the nest shown here.

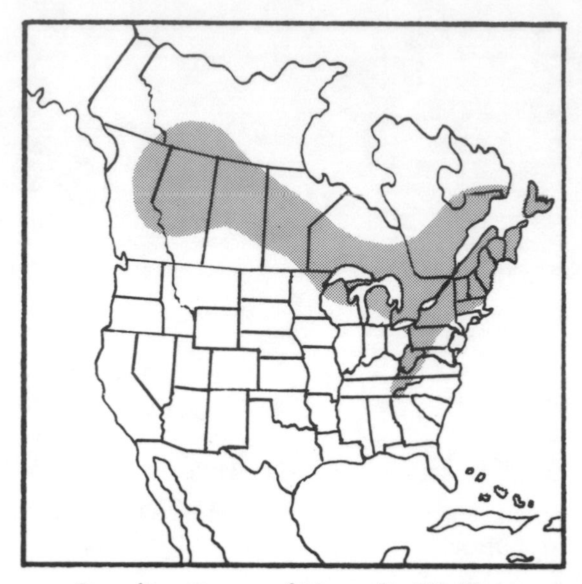

Breeding Range of Magnolia Warbler

amounts to a sign that the bird is nesting and thus has led me to many nests."[2]

Dr. Friedmann tells us that the Magnolia, "while infrequently reported as a cowbird host, is regularly victimized."[3] Over a period of nearly sixty years in southern Quebec, L. M. Terrill found 147 nests of this warbler, of which only six contained Brown-headed Cowbird eggs. My records for Maine and Pennsylvania indicate six parasitized nests in a total of thirty-one examined.

Alexander Wilson is responsible for the inappropriate species name of this warbler, *magnolia.* Of the two specimens that he secured near Fort Adams, Mississippi, in or about 1810, one was a migrant foraging in a magnolia tree. He named it *Sylvia magnolia* but gave it the English name of Black and Yellow Warbler, one that seems more appropriate than its present label.

[2] Porter, Eliot, *Birds of North America: A Personal Selection* (New York: E. P. Dutton and Company, 1972), p. 57.

[3] Friedmann, Herbert, Lloyd F. Kiff, and Stephen I. Rothstein, *A Further Contribution to the Knowledge of the Host Relations of the Parasitic Cowbirds* (Washington, D.C.: Smithsonian Contributions to Zoology, no. 235, 1977), p. 28.

15. *Cape May Warbler*

Dendroica tigrina

PLATE 6

My first expedition in search of a Cape May Warbler nest took me to Aroostook County, Maine, in 1962. After a week spent in a forest of spruces, all reaching upward in steeplelike spires, I had found only singing males. I was there in June, the right time for nesting. Females should have been laying or incubating eggs. Too late I learned of the female Cape May's clever trick for misleading predators like myself, who might follow her to the nest near the top of a tall tree. She does not fly directly to the nest; she enters the tree far below the nest and makes her way up from branch to branch near the trunk. She has another frustrating trick. When she leaves the nest, instead of flying out horizontally as most birds do, the Cape May dives toward the ground and avoids a crash with a sudden upward swoop, carrying her out of sight deceptively below the nest.

Those who have succeeded in finding nests invariably describe the site as in the uppermost clump of needles in the spirelike crown of a spruce or fir. The nest cannot be seen from the ground, even with field glasses. Richard Harlow collected seven nests of the Cape May Warbler in Tabusintac, New Brunswick, which ranged from thirty-five to sixty feet above ground. A nest I eventually found was eighteen inches from the top of a fifty-six-foot fir on Mount Desert Island.

The abundance of nesting habitat in the vast coniferous forests of the Canadian Life Zone (Coniferous Forest biome) in Canada and the northern United States would seem to afford boreal warblers such as the Cape May unlimited breeding territory. It may be surprising then to learn that millions of acres of apparently ideal sites go neglected each year. The lure to the limited areas chosen is food supply—spe-

Blake Gardner climbs to the nest of a Cape May Warbler at the top of a balsam fir on Mount Desert Island, Maine. This species places its nest in a treetop from thirty-five to sixty feet above ground.

cifically, well-documented outbreaks of the spruce budworm (*Choristoneura fumiferana* Clem), the larva of a tortricid moth.

Some researchers have suggested that the spruce budworm may be so important in the breeding cycle of the Cape May Warbler that the warblers' output of eggs, and thus of nestlings, may be correlated to numbers of this larva. Those cautious of this generalization suggest that if it is true, it would apply principally where both the Cape May and the budworm are normally most abundant, specifically northern New England and southeastern Canada. Statistics indicate that local populations of the Cape May Warbler do increase during infestations of this defoliator of coniferous forests.

Of all Wood Warblers, the Cape May's normal clutch is the largest, four to nine eggs, commonly six or seven. Other warblers that lay similarly large clutches—Bay-breasted and Tennessee—also capitalize on budworm outbreaks. These species typically lay smaller clutches during years when budworm outbreaks are few.

Betty and Powell Cottrille tell me that in Minnesota they found a Cape May's nest that held eight youngsters. After photographing the nest, the Cottrilles stood guard for two days to thwart attempts of a Broad-winged Hawk to raid the nest.

The song of the strikingly colored and patterned male is as frustrating to me as the search for a nest. It is weak, wiry, high-pitched, and beyond the range of my hearing unless the bird is close by. The notes are mainly on one pitch, a *zee* or *zeet* note uttered five or six times. I confuse it with the song of the Bay-breasted Warbler. My interpretation is *chee chee chee chee chee chee;* my wife hears it as *see see see see see.*

Outbreaks of spruce budworms in boreal forests like this lure thousands of Cape May and other warblers during nesting season.

Of all Wood Warblers, the Cape May's normal clutch is largest, four to nine eggs, commonly six or seven.

The female Cape May Warbler is more subdued in appearance than her brilliant mate. This warbler is never seen in Cape May, New Jersey, except as a migrant. Alexander Wilson named it because the first specimen was taken there.

The Cape May, a warbler that has a specialized diet during the three or four months spent in the North, surprisingly changes from an insect-eating bird to one specializing in nectar and juices sucked from fruit, which it finds abundant in its Caribbean winter home. In fact, en route south, swarms of migrating Cape Mays have descended on vineyards and caused great damage to crops. Nature has evolved for the Cape May a tubular tongue, unique among warblers, with which it efficiently exploits this winter food supply. In a decided change from its summer home in the tall spruces, the Cape May Warbler in winter is found in gardens and plantations in the mountains and lowlands of the West Indies and, infrequently, Central America.

Occasionally a bird does the unexpected. A Cape May Warbler appeared daily from December 31, 1979, to March 2, 1980, at a feeder at the home of Joe and Ruth Grom in Wexford, Pennsylvania. This bird ate peanut butter, suet, rendered fat, and some cornmeal.

A surprising prey for a Chuck-will's-widow, a close relative of the Common Nighthawk and the Whip-poor-will, was a Cape May Warbler. This bird was found nearly intact in the stomach of a Chuck found dead in Dade County, Florida, October 16, 1961. In addition,

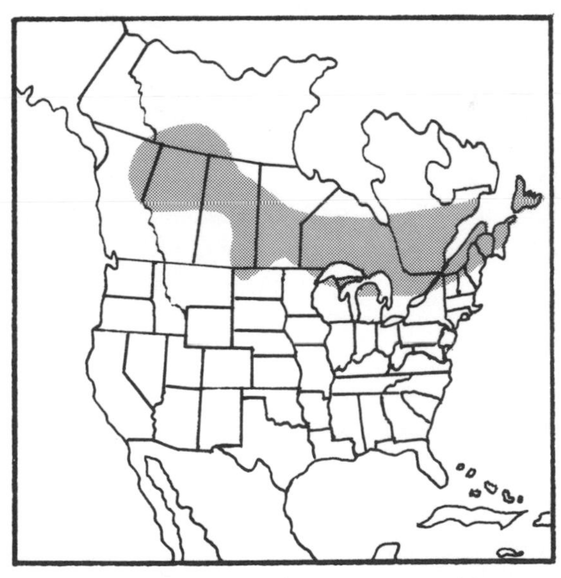

Breeding Range of Cape May Warbler

wedged tightly into the throat of the Chuck was a male Common Yellowthroat.

The Cape May Warbler was named in 1831 by Alexander Wilson, who formally described and sketched a specimen that had been collected twenty years earlier by George Ord at Cape May, New Jersey. Neither Wilson nor Audubon saw the warbler alive, and it was not reported again from Cape May until almost a century later. The species name, *tigrina* (tig-RYE-nah), is from the Latin *tigrinus,* meaning "striped like a tiger."

16. *Black-throated Blue Warbler*

Dendroica caerulescens

PLATE 4

To anyone who has puzzled over the identity of the plain, inconspicuous female Black-throated Blue Warbler, it is easy to understand why both Audubon and Wilson considered the female an entirely different species from the male. They named her Pine Swamp Warbler. Indeed, no pair of Wood Warblers is so different in appearance and so unlikely to be linked as the same species. No one has trouble identifying the distinguished blue and black male, but the female can be a problem, especially when the diagnostic white wing spot is not visible.

The preferred habitat of this species is varied. Generally, it seeks mixed conifers, hardwood forests with heavy undergrowth, and cutover areas. In Pennsylvania, I found Black-throated Blues nesting nowhere but in rhododendron bogs, especially in Cook Forest, Clarion County, where breeding pairs were so abundant that territories joined each other for miles. In the Pocono Mountains of eastern Pennsylvania, Richard F. Miller found it to be the commonest warbler where rhododendron thickets predominated. Of seventy-five nests found by Miller, approximately 90 percent were between one and three feet above ground in rhododendron.

In Maine, I have found Black-throated Blue Warblers nesting only in low conifers, mostly spruce and balsam fir. Katherine C. Harding studied fifteen nests in New Hampshire, all built in rhododendron, all between one and three feet above ground. Of forty-two nests found in mountainous areas of New York State, nineteen were in mountain laurel, thirteen in hemlock, five each in spruce and maple.

The nest of the Black-throated Blue Warbler, built almost entirely by the female, is distinguished readily from the nests of other war-

Typical habitat for the Black-throated Blue Warbler in Maine is low conifers, mostly spruce and balsam fir. Elsewhere it nests in rhododendron, mountain laurel, and hemlock.

The nest of the Black-throated Blue Warbler is distinguished from the nests of other warblers by its bulkiness and its rough exterior covered with strips and pieces of pithy wood, inner bark fibers, or birch bark.

blers by its bulkiness and its rough exterior covered with strips and pieces of pithy wood, inner bark fibers, or birch bark. Although bulky, the nest is invariably well concealed and difficult to find.

At nests I have watched, the adults are singularly quiet in the vicinity of their home. When approached, a female slips away quietly and disappears. Other observers have reported females that exhibit the "broken wing act" when approached at the nest (described in chapter 38).

Henry Mousley tells of his observations at the nest of a Black-throated Blue Warbler in Quebec where he recorded that the female fed four nestlings 349 times in fourteen hours, or once every 2.4 minutes. At the same nest, Mousley reported that in five and a half hours of watching, the female fed the young 138 times to the male's once.

The Black-throated Blue Warbler apparently rarely raises a Brown-headed Cowbird. Friedmann cites only a few records. In my own experience, I have found only one nest with a cowbird egg.

The colorful Black-throated Blue male is even more showy when he expands his wings in flight.

This fledgling Black-throated Blue Warbler will soon leave its summer home in Michigan and migrate to its winter home in the West Indies or northern South America.

A female Black-throated Blue Warbler carrying material as she constructs her nest may desert the site if disturbed. I encountered this behavior twice before I learned to move away quickly after my discovery. On one occasion, the nest was completely built and ready for eggs. I have found that the female, surprisingly, becomes quite fearless later in the breeding cycle and permits close approach while incubating or brooding. When feeding young, both adults show little or no fear of observers near by. At this time, photography is not difficult.

The song of this warbler contains the *zwee* notes characteristic of a number of other warblers, but it is husky rather than wiry. The common song goes like this: *zwee-zwee-zwee-zweeeeee,* the last note noticeably ascending on the scale. The full song is not always given. At times, only two or three notes are sung, and the trailing off at the end may be omitted.

Mr. and Mrs. Michael J. Spagnolo report a male Black-throated Blue Warbler feeding on suet at their home near Marquette, Michigan, in November, a late date for this migrant. Although there were seeds, raisins, nuts, and bread available, the warbler ate only suet. The day was snowy, with a temperature around twenty degrees Fahrenheit. Another wintered at a feeder in Kennett Square, Pennsylvania, November 8 to February 29, 1980. It survived on peanut butter and suet. These are unusual records. Ordinarily, the Black-throated Blue Warbler winters chiefly in the West Indies, but some venture farther south into northern South America, and a few linger for the winter in the southern tip of Florida.

In spring, as the birds move up the Atlantic coastal states, a race of this otherwise northern nester, known as Cairn's Warbler (*D. c. cairnsi*), ranges no farther than an area from the mountains of Georgia to Maryland. Most, however, continue farther north.

The species name, *caerulescens* (see-roo-LES-enz) is from the Latin for "blue," referring to the color of the male's crown and back.

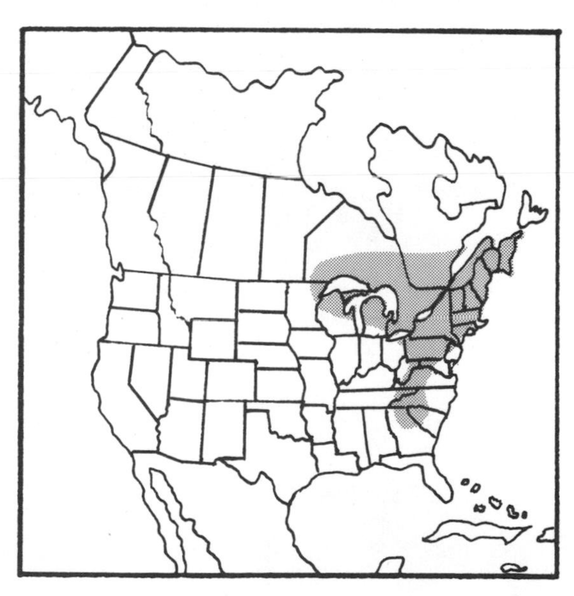

Breeding Range of Black-throated Blue Warbler

In November, 1751, a friend of George Edwards sent him a box of bird skins that he had collected aboard a ship "about eight or ten leagues distant from Hispaniola" (modern Haiti). Edwards named one of these birds the Blue Flycatcher. It later was recognized as a Black-throated Blue Warbler. It was 120 years after this discovery that the first nest of the Black-throated Blue Warbler was found in upstate New York by John Burroughs (July 1871).

17. Yellow-rumped Warbler

Dendroica coronata

PLATE 8

A decade has passed since the publication in *The Auk,* the quarterly journal of the American Ornithologists' Union, of "The Thirty-second Supplement to the American Ornithologists' Union Check-list of North American Birds," and no change made at that time has affected American bird watchers more than the "lumping" of the two well-known and common species of warblers, Myrtle (*Dendroica coronata*) and Audubon's (D. *auduboni*). That the two are conspecific was considered uncontroversial, so the English species name for the enlarged *Dendroica coronata* became the Yellow-rumped Warbler. This is the name given to the species long ago by John James Audubon.

Many bird watchers are not resigned to changing from Myrtle or Audubon's to Yellow-rumped Warbler in naming the birds in their respective ranges. There is nothing wrong with this. There are two subspecies identifiable in the field. It is not inaccurate to use the names by which the races have long been known. However, keep in mind that they interbreed readily where their ranges overlap in southeastern Alaska, eastern British Columbia, and southwestern Alberta, indicating that they and their hybrids are one species, the Yellow-rumped Warbler.

In identifying and reporting this species it seems desirable to follow the practice used in *American Birds,* a bi-monthly publication of the National Audubon Society, and elsewhere, i.e., Yellow-rumped (Myrtle) Warbler and Yellow-rumped (Audubon's) Warbler.

The evolutionary process that split *D. coronata* and *D. auduboni* from a single entity into separate geographic races undoubtedly occurred over tens of thousands of years, yet the separation and isolation apparently did not create differences so marked that two genetically incompatible species developed. The centuries of separation did, however, create differences that one might anticipate even in two so closely related subspecies.

The most obvious difference, of course, is in the appearance of the two: Audubon's has a yellow throat and the Myrtle has a white throat. Both, however, retain the four basic identifying marks: yellow on the cap, both flanks, and the rump. The species name for the Yellow-rumped Warbler, *coronata* (koh-row-NAY-tah), is from the Latin for halo, *corona,* referring to the yellow crown, a feature that the two races have in common.

The separation of breeding ranges has created some differences in feeding habits. The Myrtle gets its name from its habit of feeding on the fruit of the wax myrtle and its close relative the bayberry. (Leon A. Hausman, who made a study of the winter food of the Myrtle, reports that the waxy substance on the bayberry is not a true wax, but a fat

These fledgling Yellow-rumped (Myrtle) Warblers may not leave the United States during the winter. Many migrate no farther than the southern Atlantic Coast.

The eastern race of the Yellow-rumped Warbler (Myrtle) gets its English name from its habit of feeding on berries of wax myrtle and its close relative, shown here, the bayberry (*Myrica carolinensis*).

containing traces of protein and carbohydrates.) While Audubon's Warblers eat some fruits and seeds, their diet is about 85 percent insectivorous.

Members of both races dart into the air for insects like flycatchers, but some unusual application of this behavior by Myrtles is reported. A migrating Myrtle Warbler was observed in fall by Donald E. Kunkle following in the wake of a boat on Delaware Bay feeding on insects that it found behind the moving vessel. About thirty Yellow-rumps were observed in migration in May in Manitoba feeding on small black insects on the ice and in the water.

We attract Myrtles to our home in Florida by putting out cut-open oranges which the birds appear to relish. Numerous times I have watched one of these birds skimming across our swimming pool in swallowlike fashion, picking insects from the surface.

Aside from these minor behavioral differences, life histories of the two races of *Dendroica coronata* are so similar that what may be said of one generally is true of the other.

To me, a remarkable trait of both subspecies is the unusual way both use feathers in lining their nests, behavior unique among Wood Warblers. Shafts of feathers are woven into nest linings so that the tips curve inward over the cup, forming a screen for the eggs when the female is off the nest. All thirteen nests I found in Maine had this arrangement; in many cases the eggs were invisible from above. In one nest, after the young had fledged, I counted seventy-eight grouse feathers in the lining and canopy. All four Audubon's Warblers' nests I found in Colorado and Arizona were heavily lined with feathers. One destroyed by Steller's Jays had a thick lining of yellow, red, blue, and orange feathers.

Both races of the Yellow-rumped Warbler use feathers in an unusual way in nest linings. Shafts of feathers are woven into the lining so that the tips curve inward, forming a screen over the eggs.

Typical habitat of the Yellow-rumped (Myrtle) Warbler in Maine is a forest of conifers from four to fifty feet high.

Nests of the Yellow-rumped (Myrtle) Warbler typically are placed on horizontal branches of conifers from 4 to 50 feet above ground. My thirteen Maine nests averaged 20.5 feet; the lowest was 6 feet; the highest, 35 feet. Of forty-four nests reported by Allan Cruickshank in Maine, the highest was 43 feet above ground; the lowest, 6 feet. All were in red or white spruce or in balsam fir. Rarely, a nest is found in a hardwood: maple, apple, birch.

W. Ray Salt offers this surprising information from his experiences in Canada: "Like the Yellow Warbler, the female will often bury a cowbird egg in the base of the nest if she has not laid yet and if bothered too much will desert the nest and build again."[1] Friedmann declares that the Myrtle Warbler is a common host of the Brown-headed Cowbird in southern Canada, but that parasitization has seldom been reported elsewhere. My experience does not justify his conclusion. Five of the thirteen nests I examined on Mount Desert Island held cowbird eggs. Friedmann cites only four records of parasitization in nests of Audubon's Warblers. I found none, but my sampling was minuscule.

[1] Salt, W. Ray, *Alberta Vireos and Wood Warblers* (Alberta: Provincial Museum and Archives of Alberta Publication no. 3, 1973), p. 53.

For the western race of the Yellow-rumped (Audubon's) Warbler, open forest of ponderósa and Apache pines is a common habitat.

There is some variance in the songs of the male Yellow-rumped Warblers, but the general effect is a trill, somewhat similar to that of the Chipping Sparrow. It is my impression that the song of the Myrtle in Maine is high-pitched, not very loud, a series of short rapid notes much like *swee swee swee swee swee.* Florence Merriam Bailey considered the song's quality between that of a Yellow Warbler and a junco. Dr. Alexander Wetmore compared the song of Audubon's Warbler to that of the Myrtle with this observation: "In a way it was similar to that of the Myrtle Warbler but was louder and more decided in its character."[2]

That the Yellow-rumped Warbler reaches its breeding grounds in the East early in spring is not surprising because many may not have moved far south during the winter. No other Wood Warbler so consistently winters as far north. Reports of birds at feeding stations come from as far north as Massachusetts. It is the commonest wintering

[2] In Bent, ed., p. 268.

warbler in Florida. In the National Audubon Society's Christmas count for 1981, a total of 35,822 Yellow-rumps were counted at forty-two areas throughout the state. Topping the list was Merritt Island with 4,844; Cocoa was second with 4,192. The census was far below the 1980 count when Florida reported 86,429.

One gets a picture of the winter range of the Myrtle along the Atlantic Coast when the 1981 count is totaled: Georgia, 5,331; South Carolina, 8,012; North Carolina, 19,990; Virginia, 21,836; Maryland, 13,029; Delaware, 1,927; Connecticut 141; Rhode Island, 1,626; Massachusetts, 8,274.

The winter home of the Yellow-rumped Warbler includes the Gulf Coast and farther south into Mexico, Central America, and Panama. In March and April great flocks migrate up the Atlantic Coast, following the coastal belt of wax myrtle and bayberry. Early migration is possible largely because of this food supply. The Myrtle race is one of the few warblers that can subsist for long periods of time upon seeds and berries despite its preference for insects.

The Audubon's race is also a hardy group. They are seen during the winter throughout all but the northernmost part of their breeding range; however, some move south to Mexico and Central America, to

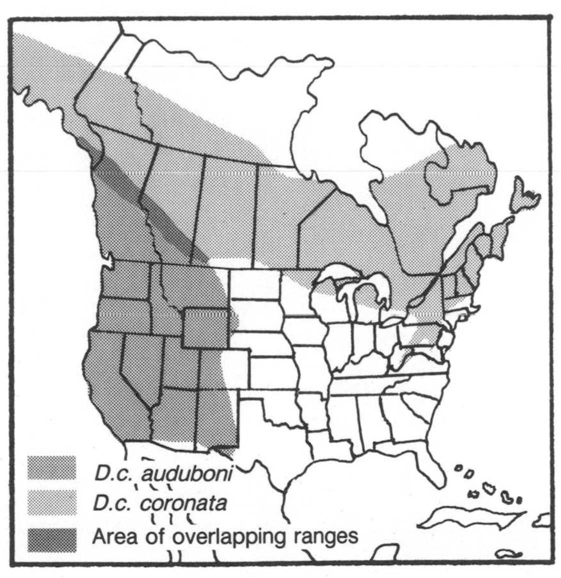

Breeding Range of Yellow-rumped Warbler

return north in spring. There is also an altitudinal migration to higher into the mountains in spring and back to lower levels in fall.

In spring, 1834, Philadelphia ornithologist John Kirk Townsend set out from Independence, Missouri, with a group of settlers and missionaries bound for the Pacific Northwest. Food became so scarce along the way that some of Townsend's bird specimens were cooked and eaten. However, Townsend preserved for his good friend John James Audubon over seventy specimens that the artist-naturalist had never seen before. One of these was a warbler new to science, secured in a forest near the mouth of the Columbia River. Townsend named it Audubon's Warbler.

18. Black-throated Gray Warbler

Dendroica nigrescens

PLATE 10

In all bird guides I have consulted, I am told that the female of this species has either a "white throat" or a "light throat" or "lacks black bib." Perhaps the females I have observed are an exceptional lot, but none had a white throat and most had as much or more black than white on the throat. At one nest where I photographed both birds, I had a problem distinguishing the sex without using field glasses. The female's black throat was flecked with white, but at a distance it looked all black (see photographs of the two birds.)

During two spring seasons in the Santa Rita and Chiricahua Mountains of southeastern Arizona, I became well acquainted with the Black-throated Gray and his almost-black-throated mate. I found seven nests, all in white or Emory oak except one in juniper. The lowest was 12 feet above ground; the highest was 40 feet; the average, 24.5 feet.

The Black-throated Gray Warbler is one of five species in the genus *Dendroica* that ornithologists believe evolved from a common prototype during the advance and retreat of the Ice Age and over a period of several hundred thousand years. The group also includes the Black-throated Green, Hermit, Townsend's, and Golden-cheeked Warblers.

Generally, each of the five occupies a separate range, but there is some overlap with Townsend's and Hermit Warblers in the northern range of the Black-throated Gray. The latter differs from the others in plumage, but also is smaller and forages and nests lower than the others. As far as I know, there is no reported hybridization between

At this nest of a Black-throated Gray Warbler it was difficult from a distance to distinguish the sexes. The male, on the left, has more black in the throat and bib; the female's black is flecked with white (below).

the Black-throated Gray and either the Hermit or the Townsend's Warbler.

The five demonstrate a relationship not only in plumage but in song. Having spent many summers in Maine where I heard the Black-throated Green Warbler sing daily, I was impressed with the similarity when I first heard the Black-throated Gray. My notes, made at the time, show my interpretation as *chee chee chee cheeteo.* Florence M. Bailey heard it about the same way when she recorded it as *zee-ee-zee-ee, ze ze ze.* As he heard it in Washington, Arthur Cleveland Bent wrote it *swee, swee, ker-swee, sick* or *swee, swee, swee, per-swee, sick.*

The Black-throated Gray Warbler winters in Mexico and casually south to Guatemala. Following spring migration it is common over much of its breeding range and may be found in quite a diversity of habitats. It is frequently found in Arizona in pinyon-juniper woodlands and in scrub oaks; in fir forests in Washington; and in manzanita thickets and chaparral in California.

Black-throated Gray Warblers are common nesting birds in the Chiricahua Mountains of southeastern Arizona. Many nests are placed in Emory and Arizona white oaks.

In one Black-throated Gray Warbler nest there were 289 feathers by actual count. Most were the soft blue feathers of the Gray-breasted Jay.

The Black-throated Gray has a knack for finding an inaccessible nesting spot on a horizontal branch, often in a fir or oak, five to forty feet above ground and four to ten feet from the trunk. Because of the inaccessibility of some of the nests I found high in oaks and far out on horizontal limbs, I was unable to check all seven. However, in one I found four warbler eggs and one Brown-headed Cowbird egg. Friedmann cited only seven cases of parasitization, although he pointed out that the warbler has been studied very little.

All the nests I have seen contained a profusion of small feathers in the lining, and I have never heard of a nest of this species that did not have feathers. One beautiful nest, destroyed by some predator, had

289 feathers by actual count. About 75 percent were the soft blue feathers of the Gray-breasted Jay.

The species name *nigrescens* (nih-GRES-enz) is from the Latin *niger,* "black," for the black in the bird's plumage. Townsend brought a specimen of the Black-throated Gray Warbler back from his western expedition.

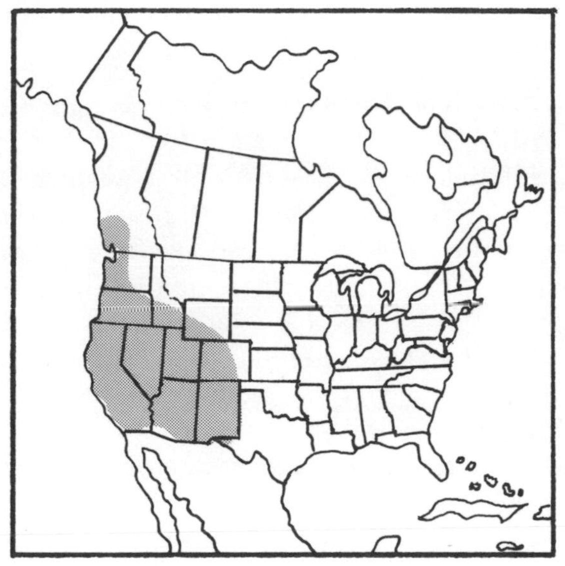

Breeding Range of Black-throated Gray Warbler

19. *Townsend's Warbler*

Dendroica townsendi

PLATE 8

In the United States, unless you are fortunate enough to catch sight of Townsend's Warbler during its migration to or from its winter home in southern California, Mexico, and Central America, look for this beautiful warbler in the vast coniferous forests of the northwestern states, Canada, or Alaska, where it nests. Perhaps the easiest way to see a Townsend's is to visit its winter home. Alexander F. Skutch, writing about the Townsend's that winter in the Sierra de Tecpan, Guatemala, declared, "They were not only the most common migratory species in the Sierra but the most numerous bird of any kind during their eight months' residence there."[1] He also observed that warblers at higher altitudes are much more gregarious than those wintering in the lowlands.

Even during May and June when nesting activities are at their peak, the birds could very easily go undetected if it were not for the male's wheezy song broadcast from the top of a tall tree. The song has been likened to that of its close relatives, the Black-throated Green and Black-throated Gray Warblers. It has the same drawling tendency and is composed of five to ten notes: *swee swee swee swee swee,* with a gradual rise in pitch toward the end.

Most of the few nests found have been in firs eight to fifteen feet above ground. The lower nests are probably exceptions and not the rule. F. R. Decker and J. H. Bowles found seven nests of Townsend's Warblers during the first twenty-eight days of June, 1923, in the Chelan region of Washington. All were in heavy woods in very rocky country; their discovery and examination involved much hard labor. According to their report, "The nests were built in fir trees from 7 to

[1] In Bent, ed., p. 288.

Townsend's Warbler nests in the vast coniferous forests of the northwestern states, Canada, and Alaska. This area along the Icicle River in the Cascade Mountains, Washington, is typical habitat.

Townsend's Warbler could very easily go undetected if it were not for the male's wheezy song.

15 feet out on the limb, and 7 to 60 feet above ground. Each nest was directly on the limb under a protecting spray of needles. The eggs were four or five in number. In each instance, the female allowed a close approach, but then dropped straight to the ground and disappeared. The males were not seen in the vicinity."[2]

The only record of parasitization of Townsend's by the Brown-headed Cowbird seems to be one reported by Dr. J. B. Tatum in southern Vancouver Island. The report involved a fledged cowbird attended by its foster parents.

Examinations of many specimens of this warbler indicate that insects and spiders make up about 95 percent of the birds' diet. Stinkbugs are particular favorites. However, in winter, Townsend's have been attracted to feeding stations along the Pacific coast, where they have been watched eating cheese, marshmallows, and peanut butter.

Townsend's Warbler was named for John Kirk Townsend, who collected the first specimen of the bird in 1837 in "the forests of the Columbia River," probably what is now the state of Washington.

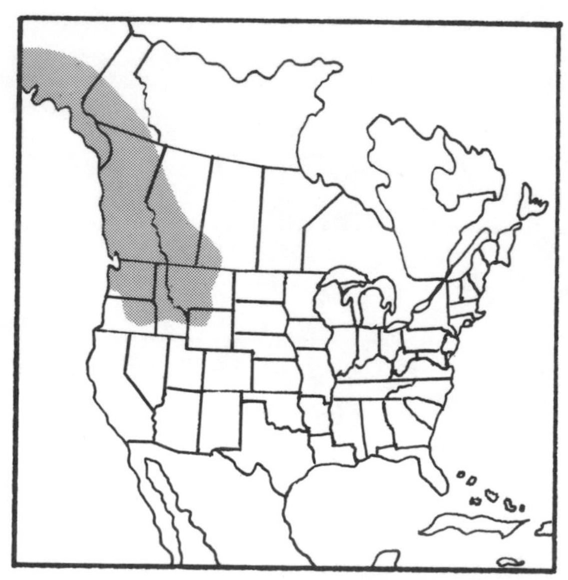

Breeding Range of Townsend's Warbler

[2] Decker, F. R., and J. H. Bowles, "Bird Notes from Chelan County, Washington," *Murrelet,* vol. 4, p. 16, 1923.

20. *Hermit Warbler*

Dendroica occidentalis

PLATE 10

What has been said of Townsend's Warbler in the preceding chapter is generally true of the Hermit Warbler, its close relative. Like the Townsend's, it frequents heavy timber, especially second growth, and is difficult or sometimes impossible to find except when the male sings during May and June. It lives in the crowns of coniferous trees, dropping down only occasionally. To bird watchers, it is one of the most wanted of western warblers for life lists. For those who want to bring the Hermit Warbler into closer range, a suggestion has been made by H. L. Cogswell. "In the Sierra Nevada of California I have found that even the males can be called down to eye-level by a patient and quiet observer imitating the call of a Saw-whet Owl."[1] Once seen, the adult male Hermit Warbler with his triangular black cravat and yellow face is almost unmistakable.

The birds arrive from their winter home in Mexico and Central America in early May to establish territories in the tall Douglas fir forests. Arrival is detected first because of the male's song, which has been translated as *zeegle-zeegle zeegle-zeek,* uttered somewhat slowly at first but ending rather sharply. It is a penetrating twitter, harsher and more run together than the song of the Chipping Sparrow. Chester Barlow described it this way: *tsit tsit tsit tsit chee chee chee,* the first four syllables gradual and of uniform speed, ending quickly with *chee chee chee.*

[1] Griscom, Ludlow, and Alexander Sprunt, Jr., *The Warblers of America,* (New York: The Devin-Adair Company, 1957), p. 144.

The tall pines of western Washington are an example of ideal breeding habitat for the Hermit Warbler.

The Hermit Warbler is one of the most wanted by life-listers. It may be lured into closer range by use of the call of the Northern Saw-whet Owl.

It is said that no other species is so unswervingly regular in its nesting habits. You must find the nest between June 3 and June 15 if you want to see eggs. The few nests of these species that have been found have usually been within fifteen feet of the ground, but it is known that most Hermit Warblers nest one hundred feet or more above the ground in giant conifers.

The nest is saddled to a horizontal limb. It is built of forb stems, pine needles, and woody plant fibers held together by spider webs.

The Hermit closely resembles Townsend's Warbler, and two well-defined hybrids have been collected in Arizona. Stanley G. Jewett reported a number of hybrids from Washington and Oregon. The hybridization has been studied extensively by Michael L. Morrison, who presented a paper on the subject at the 1983 meeting of the Cooper Ornithological Society. According to Morrison, the ranges of the two species meet but do not overlap broadly. The Hermit Warbler is not known to hybridize with the Black-throated Gray Warbler, although they are closely related and their breeding ranges also overlap.

The nest of the Hermit Warbler is saddled to a horizontal limb of a giant conifer, sometimes one hundred feet or more above ground.

The Brown-headed Cowbird does not seem to be a factor in the nesting life of the Hermit. Friedmann records only one instance of parasitization. A fledgling cowbird was seen being fed by a pair of Hermit Warblers at Camp Augusta, near Nevada City, California, on June 21, 1942.

John Kirk Townsend described the Hermit Warbler in 1837 from a pair taken at Fort William (now Portland), Columbia River, May 28, 1835. The species name, *occidentalis* (ock-sih-den-TAL-lis), is Latin for "western," from *occidere,* "to fall or to set (as the sun)." Hermit suggests the warbler's shy and retiring habits in the great coniferous forests of the Pacific states.

Breeding Range of Hermit Warbler

21. *Black-throated Green Warbler*

Dendroica virens

PLATE 11

In my study of the songs of Wood Warblers, it has become apparent to me that the song of the Black-throated Green has been described by more would-be interpreters in more different ways than the voices of any other members of the family.

The bird is abundant in its chosen habitat, and the male is a prolific songster. He drawls his lazy spring song from high in the thick branches of spruce, fir, pine, and hemlock and is very much a characteristic part of the northern woods.

But what does he say? My ears hear one song as *dee-dee-dee-de-ert.* Standing right beside me, my wife hears *zree-zree-zree-zer-zreet.* The first three syllables are on one pitch; the fourth a tone lower; and the fifth the same as the first three. Here are some other interpretations: To O. W. Knight, "the typical song is rather jerky and sounds like *te-he-think-o-me*"; to C. J. Maynard, it sounded like, "*Good Saint The-re-sa*"; Margaret Morse Nice heard two main songs, "*Trees, trees, murmuring trees*" and "*see see see wee see*"; and Bradford Torrey heard, "*Sleep, sleep, pretty one, sleep.*" About these, Kenneth C. Parkes comments, "This is a mélange of descriptions of the two very distinctive songs. Mrs. Nice's descriptions, which distinguish between the two, are the best."[1]

The male sings one pattern when near the nest or in the presence of the female, and another at the boundaries of his territory or in

[1] Personal correspondence.

The Black-throated Green Warbler is abundant in its habitat of open evergreen or mixed forests. Less frequently a pair chooses a hardwood stand of birch, beech, or poplar.

confrontation with another male. Unmated males have been heard singing only the latter variation.

In 1931, Mrs. Nice made a definitive study of the Black-throated Green Warbler in Pelham, Massachusetts. During ninety-four hours of observation on fifteen different days in July, Mrs. Nice counted 14,005 songs rendered by a single male. The average for the period was 149 songs per hour, but there was one hour on July 15 when the male sang 466 individual songs. The longest period of uninterrupted singing was seventy-four minutes.

The Black-throated Green is one of the best-known Wood Warblers among the great waves of migrants that move north in spring. During late April and May, the migrant army reaches its numerical peak, and by June Black-throated Greens are scattered throughout their breeding territories. They remain until August or September, then return south to their winter headquarters in Florida (sparingly) and southern Texas south to Colombia.

On arrival in the North, the Black-throated Green Warblers head for rather open forests of evergreens. Less frequently a pair chooses a hardwood stand, especially birch, beech, and poplar. I have found nests only three feet above ground in sapling spruces and, at other times, as high as forty feet on horizontal limbs of conifers. The average is about twenty feet.

Where there are birches near, the female is prone to add papery strips of white birch bark to the outside walls of the nest, little white curls that flutter in a breeze and make the nest easier to find in dark coniferous forests. Typically, the nest is a compact, well-built, deep cup saddled to a branch or fork of a conifer, or sometimes a hardwood.

Whether the male feeds the young while they are in the nest seems to vary with individuals. One season in Maine, I photographed two nests. At one, the male never fed. He never came near the nest when I was watching. He sang only at a distance. At the second nest, the male fed almost as often as the female. Frank A. Pitelka observed

The female Black-throated Green Warbler adds papery strips of white birch bark, when available, to the outside walls of her nest. The nest is commonly saddled to the horizontal limb of a conifer.

The nest of the Black-throated Green is built by the female from three to forty feet above ground. The average is about twenty feet.

that both sexes feed the young, the female more often. He added, "At almost every one of his visits the male, although feeding less frequently, brought more food than the female."[2] In Mrs. Nice's experience, in sixteen hours at one nest she saw forty-six feedings by the female and none by the male.

On September 7, a surprisingly late date for such an observation, Mrs. Nice saw a molting Black-throated Green Warbler feeding two fully grown young out of the nest. Her previous records were August 21 and 23.

Friedmann in 1963 reported only fifteen known incidents of cowbirds parasitizing nests of Black-throated Green Warblers and declared it "a very infrequent victim," but by 1977 eighteen more cases

[2] Pitelka, Frank A., "Breeding Behavior of the Black-throated Green Warbler," *Wilson Bulletin,* vol. 52, no. 1 (1940), p. 8.

were added to his records. Of eighteen nests I examined on Mount Desert Island, two were parasitized.

It may come as a surprise to learn that a race of the Black-throated Green Warbler commonly known as Wayne's Warbler (*D. v. waynei*), is a summer resident in the coastal regions of southeastern Virginia, and North and South Carolina. It was named for Arthur T. Wayne of Bachman's Warbler fame, who discovered this subspecies. Like its northern counterpart, this southern subspecies is migratory, but Brooke Meanley tells me that nests have been found as early as April 4, long before the northern birds arrive on their breeding territories.

Young Black-throated Green Warblers are in the nest eight or nine days before they leave. The fledglings here are still dependent upon their parents for food.

He has seen the birds in the Dismal Swamp of Virginia and North Carolina as early as late March.

The species name, *virens* (VIR-enz) is from the Latin *virere* meaning "to be green," and refers to the bird's greenish back. Johann Gmelin, who followed in the footsteps of Linnaeus, named the bird in 1789, using a description from later discoveries by George Edwards.

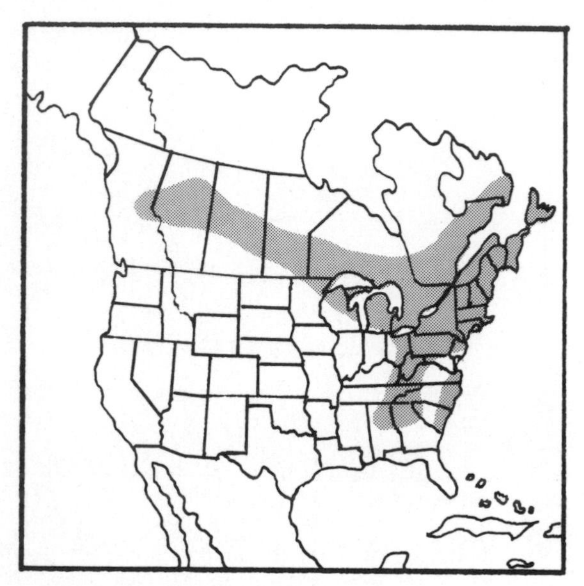

Breeding Range of Black-throated Green Warbler

22. *Golden-cheeked Warbler*

Dendroica chrysoparia

PLATE 10

Like Kirtland's Warbler, which nests only in Michigan and only in stands of young jack pine, the Golden-cheeked Warbler has very specific requirements and nests only in Texas and only in areas of mature Ashe juniper. According to Warren M. Pulich, author of the definitive book, *The Golden-cheeked Warbler,* these warblers have been reported only twice anywhere in the United States outside of Texas, a bird in Florida and another on the Farallon Islands off the California coast. The Tropical Parula and Colima Warbler are also confined to Texas within the United States, but these both breed extensively in Mexico as well.

There are other species of juniper in Texas, but only the mature Ashe juniper meets the needs of the rare Golden-cheeked Warbler. Ashe juniper is common in rough rocky slopes and canyons of the Edwards Plateau in central Texas and northward, where the Golden-cheeked has been recorded in forty-one counties. At present, the species breeds in only about three-quarters that number of counties and typically in small pockets of "cedar," as the junipers are called locally.

The reason the Golden-cheeked Warbler nests only in these mature stands of Ashe juniper, or "cedar brakes," seems to be the female's demand for long strips of bark from mature Ashe juniper for construction of her nest. The nest may or may not be in juniper; but, regardless of where it is placed, these bark strips are invariably part of its structure.

The nest, built entirely by the female, is in a tree fork five to thirty feet above ground. The bark strips used in the outer wall blending with the bark of the nest tree offer artful camouflage.

It was during nesting season that I first ventured into Golden-cheeked Warbler country. Males of all species were singing, but I knew very quickly that I was about to see a Golden-cheeked. I could hear a hurried buzzing that sounded much like the song of the Black-throated Green Warbler, which I know well. Frank Chapman described the Golden-cheeked Warbler's song as *tweah, tweah, twee-sy.* Peterson heard it as a hurried *bzzz, layzeee, dayzeee.*

The Golden-cheeked is one warbler whose scientific name is appropriate. The species name *chrysoparia* (kris-op-AY-rih-ah) is from the Greek *chrysos,* "golden," and *pareia,* "cheek."

Regarding the future of this rare bird, Warren Pulich states: "One of the serious problems facing the Golden-cheeked Warbler is its loss of habitat. . . . Because the species has a narrow tolerance for habitat

View along the Pedernales River in Blanco County, Texas, where the Golden-cheeked Warbler finds the mature stands of Ashe juniper it demands for breeding.

A mature Ashe juniper, which supplies the female Golden-cheeked Warbler the long strips of bark she requires for constructing her nest in the "cedar brakes" of Texas.

requirements, a particular population is eliminated whenever a cedar brake is destroyed. One cannot leave a few old junipers and expect the species to remain. . . . There must be careful management planning for this species. . . . At the present rate of clearing, juniper suitable for nesting birds may be eliminated by the turn of the century."[1]

In one instance at least I can testify to the fact that Pulich's fears are well founded. In April 1973, when I visited Pedernales State Park, I found the Golden-cheeked Warblers particularly abundant in an area overlooking the river. When I returned a few years later, this area had been converted into a group campground.

For bird watchers anxious to add this species to their life lists, several areas are suggested: near Dallas in Meridian State Park; at Austin, the Travis County Sanctuary or Austin City Park; and at San An-

[1] Pulich, Warren M., "The Golden-cheeked Warbler" (Austin: Texas Parks and Wildlife Department, 1976).

tonio, a wilderness park, twelve miles north of Loop 410 on I-10. I have been successful in finding the birds nesting in the cedar brakes at Pedernales Falls State Park, Blanco County. The park is near Johnson City on U.S. Route 290.

San Antonio is where the first Golden-cheeked Warbler was recorded for the United States. It was taken in about 1864 by a man named Dresser. The first specimens ever recorded were taken in Guatemala.

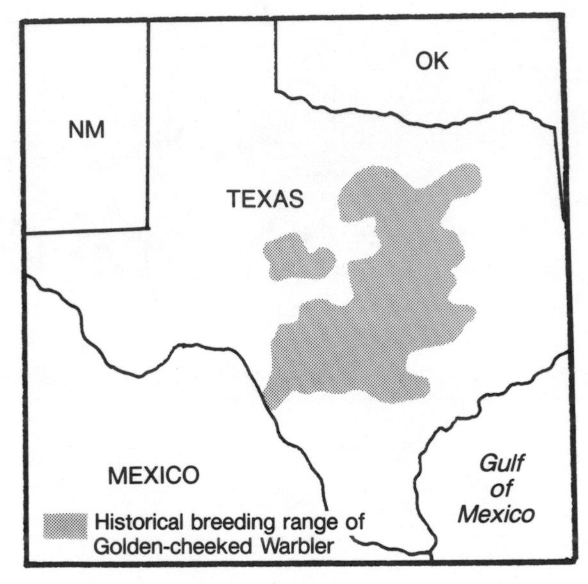

Breeding Range of Golden-cheeked Warbler

23. *Blackburnian Warbler*

Dendroica fusca

PLATE 14

One of the most delightful experiences in bird watching is to see the first male Blackburnian Warbler each spring, with his flaming orange head and throat accentuated brilliantly against the deep green of a coniferous forest. For me, it is an annual event that never loses its thrill or its excitement through repetition. In late summer or early fall when I last saw this gaudy bird winging southward toward its winter home in Central America and northern South America, it was wearing more somber plumage, a far cry from the flaming feathers acquired before its return.

In spring the Blackburnian Warbler moves north to its breeding grounds, where it will be at home for four or five months in mature conifers—hemlocks, pines, firs, and spruces—and in mixed forests. In the southern part of its breeding range, the Blackburnian nests in places where conifers are scarce or absent. Here it has adapted to mature stands of deciduous trees, often oaks. In suitable habitat, the birds may congregate in loose colonies, but a typical pair will generally establish a nesting territory uninfluenced by others of its kind.

I have learned that Blackburnian Warbler nests are exceedingly hard to find. They are tucked into dense branches of conifers and also placed high in the trees, where they are usually invisible from the ground. Of seven nests found by Allan Cruickshank in Lincoln County, Maine, the lowest was forty-three feet above ground, the highest, seventy-six feet.

The female Blackburnian builds the nest alone in three to six days. I have watched one gather grass and small forb stems on the ground and carry the material to a small fork near a treetop. Or the nest may

Several singing males could be heard along this spruce-lined back road in Maine, ideal nesting habitat for Blackburnian Warblers.

be saddled on a horizontal limb, usually far from the trunk. The nest is made of twigs, plant down, usnea when available, and spider silk, with fine grass included in the lining.

In a Maine white spruce tree at a nest that contained two Blackburnian Warbler nestlings and one very large Brown-headed Cowbird youngster, I attempted to get color photographs of the adult male. During each visit by the male, the cowbird received all the food he brought and dominated the photograph. For a better view of the Blackburnian, I removed the cowbird, leaving the two small warbler nestlings. The male appeared a few minutes later, his bill full of food. He apparently missed the cowbird for he seemed to be searching the nest. He then flew away without feeding his own offspring, and never returned to the nest during the two days that I observed it, although he remained in the territory. The disappearance of the cowbird apparently triggered his desertion. The female continued to feed, and the young warblers fledged two days later.

Considering the inaccessibility of the Blackburnian's nest, it is not surprising that it is not often reported to be a victim of the Brown-headed Cowbird. One of the few records was a discovery by C. H. Merriam of a Blackburnian's nest eighty-four feet above ground containing four warbler eggs and one of the cowbird. (It takes a heap of climbing to study Blackburnian nests.)

Nests of the Blackburnian Warbler are tucked into dense branches of conifers high in the trees, usually invisible from the ground. Spruce twigs in the outer walls blend with the site.

Blackburnian Warblers are not common victims of the cowbird, but this nest did not escape. The male is feeding a nestling cowbird almost as large as himself.

The female Blackburnian Warbler is so much paler than the male that Audubon and Wilson thought it another species. This female has come to a woodland pool for a drink and a bath.

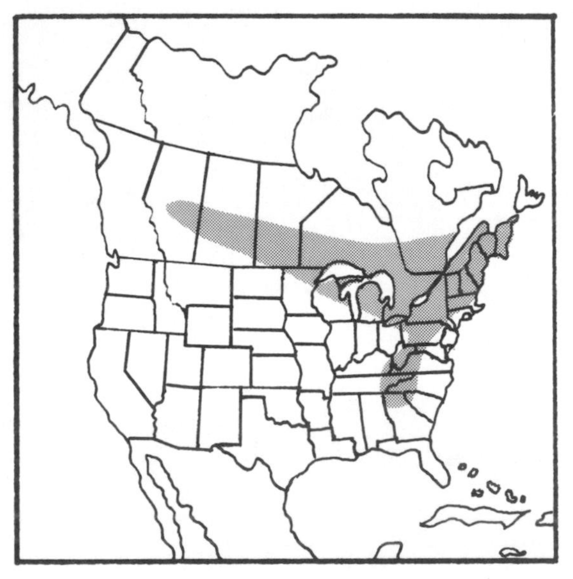

Breeding Range of Blackburnian Warbler

The beauty of the male is undeniable; his song is at best disappointing. It is a weak, thin, high-pitched trill ending in a wiry note that is too high for many of us to hear. The song isn't loud, but to many it is penetrating. According to Brewster, the song is somewhat like that of the Northern Parula and may be represented by the syllables *tseep, tseep, tseep, tse-tse-tse,* with the last three notes on a higher pitch than the first.

From the name Blackburnian, one might deduce that "black" and "burn" referred to the bird's orange and black plumage. Not so! It was named in 1788 by Johann Gmelin for Anna Blackburne, an English botanist. Gmelin called the species *Motacilla blackburniae.* The name *blackburniae* was later discarded because of a prior specific name, *fusca,* given by P. L. S. Muller, a Dutch professor. *Fusca* (FUSS-kah) is Latin for "dark" or "dusky." The English name has, however, been retained.

The female Blackburnian Warbler is so much paler than the male that Audubon and Wilson thought it a different species. They named it Hemlock Warbler, a name appropriate only in the southernmost part of the bird's breeding range.

24. *Yellow-throated Warbler*

Dendroica dominica

PLATE 13

Two days spent with my son in Westmoreland State Park in Virginia studying and photographing the Yellow-throated Warbler left me with two lasting impressions of this bird: its characteristic song, which reminded me of a Louisiana Waterthrush; and its unique habit of creeping along limbs and trunks of trees as it foraged for insects. No other warbler except the Black-and-white compares in its ability to search over vertical surfaces. As I learned later, the Yellow-throated has the

The author comes face-to-tail with a male Yellow-throated Warbler feeding young in a nest in Westmoreland State Park, Virginia.

It took a long extension ladder to reach this nest of a Yellow-throated Warbler, forty-five feet above ground, near Tallahassee, Florida. This is typical habitat.

longest bill in the genus *Dendroica,* presumably an adaptation for probing into bark.

This creeping habit did not escape the observations of early American explorers of natural history. Mark Catesby, who first came to the United States from England in 1710, called the species the Yellow-throated Creeper. This method of feeding continues into the winter among those that migrate to Central America and the West Indies. Kenneth C. Parkes tells me that he has seen it in winter in Yucatan creeping across the fronts of main-street buildings, searching for flies.

The Yellow-throated Warbler can be found year-round from central Florida north to South Carolina, although those that winter in the area are not necessarily the birds that nest there. There is a general migration southward, and some individuals move south in autumn as far as Central America and the West Indies. In fact, the specific name of the warbler, *dominica* (dom-IN-ih-kah) refers to Santo Domingo, an early name for Hispaniola, where the first specimen was collected.

In the southeastern United States, especially coastal regions of the Carolinas and south, Yellow-throated Warblers are so attracted to Spanish moss as a nesting site that it is almost useless to search for them elsewhere. Their close association with strands of the air plant is comparable to that of Northern Parulas. Inland, where Spanish moss is not available, Yellow-throateds nest high on horizontal limbs of pines where their nests are concealed in pine needles.

In Spanish moss, the female forms a pocket and weaves into it grasses and additional strands of moss to form a nest cup. Open nests are built of bark strips, grasses, and plant down, and often lined with

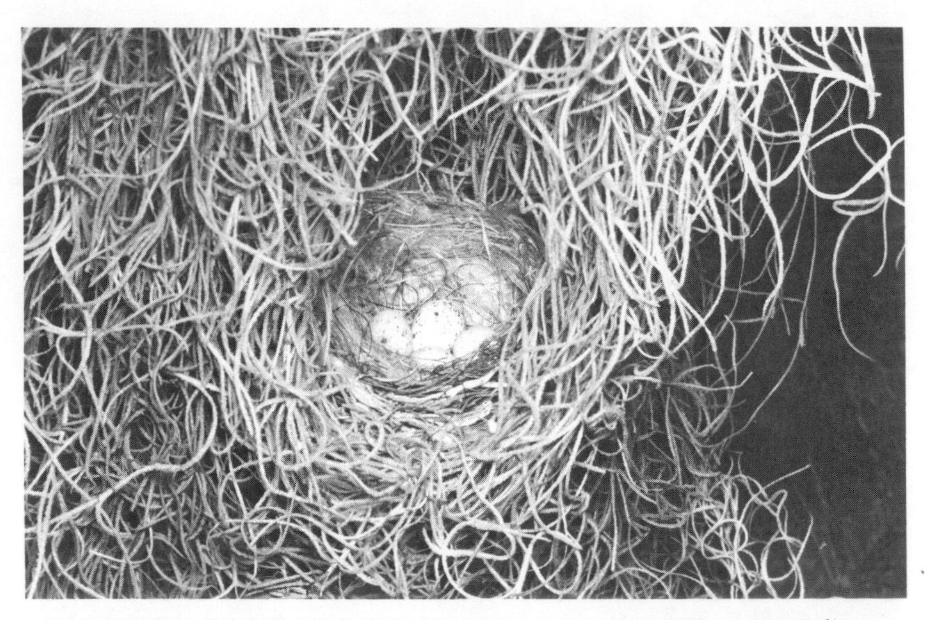

In the South, the Yellow-throated Warbler is so attracted to Spanish moss as a nesting site that it is almost useless to hunt for it elsewhere.

Male Yellow Warbler

Kentucky Warbler

Male Black-throated Gray Warbler

Male Hermit Warbler

Female Golden-cheeked Warbler

Black-throated Green Warblers, male, top; female, bottom

Kirtland's Warbler (endangered species)

Prairie Warbler

Palm Warbler pair, male feeding female on nest

Grace's Warbler

Male Yellow-throated Warbler

Pine Warblers,
male, right; female, left

Male Blackburnian
Warbler

Blackpoll Warbler

PLATE 14

Bay-breasted Warblers, female, right; male, left

Adult male
American Redstart

First-year male
American Redstart

Female
American Redstart

This Yellow-throated Warbler had a very successful nesting in coastal Virginia: from five eggs came five healthy fledglings.

feathers. Nests have been discovered as high as one hundred feet above ground, but the average is probably near thirty feet.

The song of this species has been compared to those of the Indigo Bunting, Carolina Wren, Prothonotary Warbler, and Louisiana Waterthrush. It has always reminded me of the waterthrush's song because of its loud, carrying quality.

A western subspecies makes its summer home west of the Appalachian Mountains, principally in the Mississippi Valley. It is known as the Sycamore Yellow-throated Warbler (*D. d. albilora*) because it is partial to sycamore trees for nesting. In recent years, this migratory subspecies has been extending its range northward. As recently as 1983, the bird's range in western Pennsylvania was expanded northward to Armstrong County. A nest was found in Frick Park in the city of Pittsburgh.

Friedmann's only report of cowbird parasitization is from Oklahoma and involves this subspecies.

My association with a small party of tenacious bird watchers who call themselves the Sutton Searchers has been one of the outstanding

A painting by George M. Sutton of a pair of Sutton's Warblers, believed to be hybrids between a Northern Parula and a Yellow-throated Warbler.

memories of my years of watching warblers. This group is an offshoot of the Brooks Bird Club, of Wheeling, West Virginia. It was formed in 1951 for the purpose of searching in May each year for a hybrid bird, Sutton's Warbler, discovered in the eastern panhandle of West Virginia in 1939 by Karl W. Haller and Lloyd Poland.

The hybrid is believed to be a cross between a Yellow-throated Warbler and a Northern Parula. It has been given the English name Sutton's Warbler (for Dr. George M. Sutton) and the scientific name

Dendroica potomac. Only two specimens, a male and a female, have been collected. A few sight records have been reported but none by the Sutton Searchers. Despite their failure ever to see or hear Sutton's Warbler, members of this group have shown unbelievable persistence and determination. They have met in Harper's Ferry in late May for the past thirty-three years, where they spend several days searching the area where Haller and Poland collected the two hybrids.

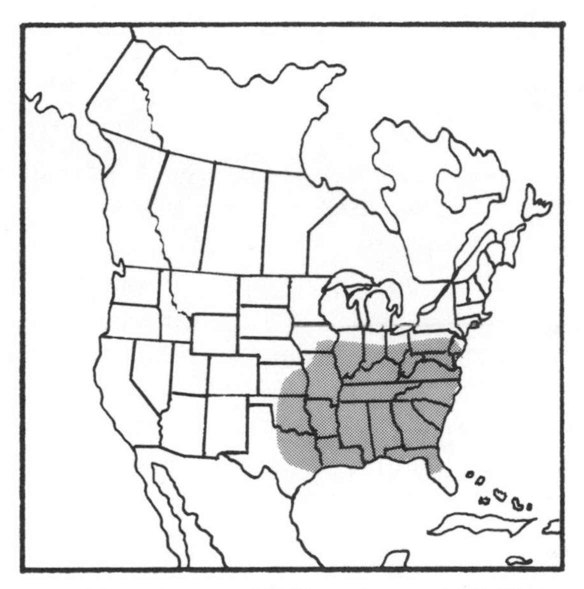

Breeding Range of Yellow-throated Warbler

25. Grace's Warbler

Dendroica graciae

PLATE 13

Allan R. Phillips and J. Dan Webster, outstanding western ornithologists, have declared that "Grace's Warbler (*Dendroica graciae Baird*) can probably claim the dubious distinction of being the least understood of all the widespread North American wood warblers."[1]

In view of that statement, I felt fortunate to observe Grace's Warbler at six different nests in the Chiricahua Mountains of southeastern Arizona in May and June, 1979. The nests were all forty to sixty feet above ground in ponderosa and Apache pines; and all were in clumps of needles near the ends of horizontal limbs except two that were in the crowns of pines.

Four nests that were accessible contained eggs or young. All the nests were very much alike, very compact and somewhat flat. The bases and sides were composed mainly of oak catkins gathered from nearby trees. They were lined with hair, fine grasses, rootlets, and feathers. The use of catkins gave a spongy feel to the nests.

At the sites where I watched construction, the female alone gathered and placed material. On a few occasions the male accompanied her, but never closely. He usually sang nearby. In approaching the nest, the female flew into the tree at a place away from the nest site and worked her way out from the trunk, creeping slowly along the nest branch to the selected site, which was hidden in a dense clump of needles at the tip of the branch.

Grace's Warbler, a close relative of the Yellow-throated Warbler, is

[1] Phillips, Allan R., and J. Dan Webster, "Grace's Warbler in Mexico," *The Auk,* vol. 78, no. 4 (1961), p. 551.

A bird of the tall pines, Grace's Warbler seldom comes to the ground. An extension ladder was used to examine a nest in Barfoot Park in the Chiricahua Mountains, Arizona.

One of six nests of Grace's Warbler studied by the author in the Chiricahua Mountains. All were forty to sixty feet above ground in pine. This nest was placed deep into the shelter of long Apache pine needles.

Grace's Warbler spends the summer in pine-oak forests of western mountains, where it nests at altitudes typically above six thousand feet.

a tiny denizen of the great pine-oak forests of western mountains, where it nests at altitudes typically above six thousand feet. It is a bird of tall pines, seldom coming to the ground. Webster, who has made an extensive study of the four races of this species, states that "the predilection of Grace's Warbler for pines is extreme; only twice have I seen one land in an oak tree, and then it was nervous and not foraging."[2]

The male Grace's Warbler sings throughout his territory, usually from high in a pine tree. The song is a single note repeated rapidly, *chip chip chip chip chip,* and given in a loud tone with the last syllable as strongly accented as any of the others. He assists in feeding the young. Both birds dart toward the nest tree very rapidly, and both use the creeping approach on the nest limb after reaching the trunk. Departure from the nest is so swift that an observer can miss it while blinking his eyes.

Cowbirds don't seem to share human problems in finding Grace's Warbler nests. Friedmann reports several cases of parasitization.

[2] Webster, J. Dan, "A Revision of Grace's Warbler," *The Auk,* vol. 78, no. 4 (1961), p. 556.

Grace's is an active bird, restless and constantly on the move. During song periods, the male flies rapidly from one favorite perch to another. His flight is quick and jerky. The species has a habit of catching some of its prey on the wing. I have also observed a Grace's Warbler hovering over pine needle clusters and cones in search of insects.

Following the nesting season, the birds retire to their winter home in Mexico and Central America, where two southern races of Grace's Warbler are permanent residents.

The first specimen was collected by Dr. Elliott Coues, a surgeon-naturalist in the United States Army at Fort Whipple, near Prescott, Arizona, in 1864. It was named for Dr. Coues's sister, Grace Darling Coues, in 1865 by Spencer F. Baird, assistant director of the Smithsonian Institution. The first nest was found in 1890 by H. P. Attwater in Yavapai County, Arizona.

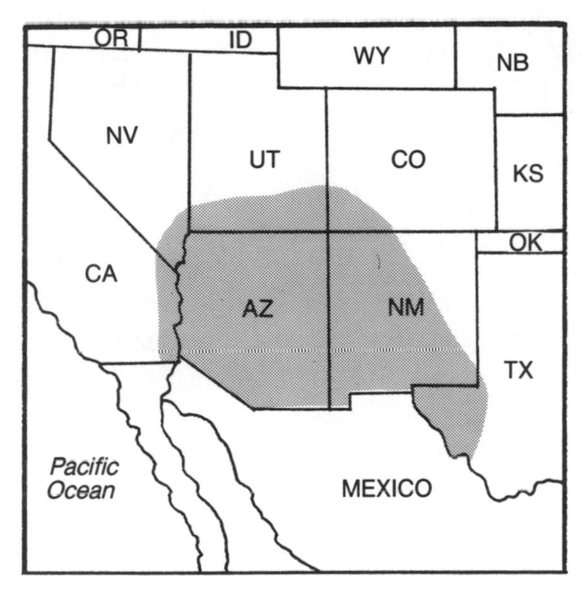

Breeding Range of Grace's Warbler

26. *Pine Warbler*

Dendroica pinus

PLATE 14

Even we who live in the South can boast of very few Wood Warblers that are with us for the entire year, but the Pine Warbler is one of them. In winter, when few warblers are seen in the North, we have an abundance of Pine Warblers. Those that nest in the North have a comparatively short migration, for they move only into the Carolinas, south to Florida, and across the Gulf states to eastern Texas. In the spring and summer, when many birds have gone north again, our pine woods ring with the musical high-pitched trills of this inconspicuous bird.

The name Pine-creeping Warbler, given to this species by both Wilson and Audubon, is indicative of the bird's habitat and its behavior. Much like a Brown Creeper, a Nuthatch, or a Black-and-white Warbler, the Pine habitually creeps over tree trunks and limbs in search of hidden eggs or larvae. Although normally quite deliberate in its movements, it is adept at striking out in flycatcher fashion for flying insects.

The Pine Warbler is rarely found far from open pine woods, especially during the breeding season. Although nests have been discovered in cedar and cypress, this is unusual. Fifteen species of trees in the genus *Pinus* are known to have been used. Nests are invariably difficult to examine for they are either placed high in pines, 25 to 40 feet above ground, or very far out on horizontal limbs or both. They are rarely lower, often higher; a South Carolina nest 135 feet above ground holds the record. Of eight nests that I have found, only three were accessible. Usually the nest is completely hidden from below by a cluster of leaves that surrounds it.

The Pine Warbler is rarely found far from open pine woods, especially during breeding season. Nests are built high in the trees and typically far out on horizontal limbs.

In the South, nesting begins early, and it is possible that three broods may be raised annually. Fresh eggs have been found as early as March 15 in Georgia. In the North, nesting usually begins in May.

In Georgia in March a female Pine Warbler was seen gathering nesting material from the side of a crow's nest that held eggs but from which the owner was temporarily absent. Another female observed a few days later was engaged in lining her nest with bunches of hair plucked from a dead cat lying in an open field. At State College, Pennsylvania, I watched a female Pine Warbler carry material from a nest she had deserted to one in a new location nearby.

Friedmann calls the Pine Warbler a rarely reported victim of the Brown-headed Cowbird. Only twelve instances are known.

Arthur Cleveland Bent states that "several observers have noted that the male shares with the female the duties of incubation."[1] I have no evidence for that statement, but I have watched both adults feed the young.

This warbler normally is an insectivorous bird, but in winter it will come to feeding stations for bread, cornmeal, peanut butter, nutmeats

[1] Bent, p. 410.

Of eight Pine Warbler nests that I have found, only three were accessible. This one in Duval County, Florida, was thirty feet above ground but unusually close to the trunk.

Most Pine Warblers remain in the United States throughout the winter. In the South, they nest as early as mid-March and raise two or possibly three broods annually.

and especially beef suet. It will also feed on berries such as sumac, ivy, grape, and bayberry.

Edward Howe Forbush told about an incident in Springfield, Massachusetts, when a young Pine Warbler kept inside a schoolroom for three days was fed constantly by its parents. The adults flew in through a window even though the school was in session.

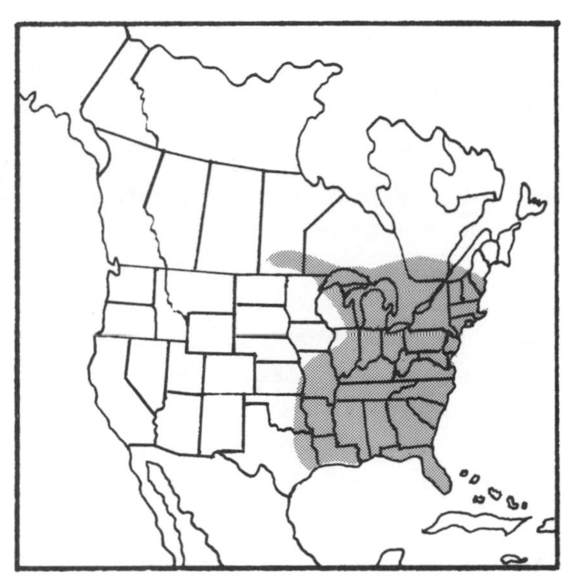

Breeding Range of Pine Warbler

27. *Kirtland's Warbler*

Dendroica kirtlandii

PLATE 12

No Wood Warbler, and possibly no other North American bird, has received as much media attention and public interest as the endangered Kirtland's Warbler. Annually, newspaper, magazine, and television coverage is given to the nesting success or failure of this species in its limited breeding area in about five or six counties in the northern half of Michigan's lower peninsula.

In the 1950s, when my son and I went to Michigan in search of Kirtland's Warbler, no measures were taken to protect this rare and endangered bird from the public. We were free to tramp the jack pine forests without any restrictions. Fortunately, the area is now closed to the public from May 1 to August 15. Guided tours are conducted to permit birders to see the warblers, but under controlled conditions.

The United States Forest Service, the United States Fish and Wildlife Service, the Michigan Department of Natural Resources, and the Michigan Audubon Society are involved in management efforts. A Kirtland's Warbler Recovery Team heads the project. An area of 135,000 acres of jack pine habitat has been designated for attention in the Kirtland's Warbler's historical breeding range in Michigan. The eventual goal is to increase the population to one thousand pairs, a goal that seems far away at present.

It has been estimated that the total weight of all the Kirtland's Warblers in the world would be less than twelve pounds, for an adult weighs only one-half ounce. In the entire history of this warbler, all the nests found have been in only thirteen adjacent counties. However, in 1982 and in 1983 a singing male was heard for the first time in the Upper Peninsula near Gwinn. Singing males, but no females and

Kirtland's Warbler demands large stands of jack pine six to twenty feet tall and porous soil with low, dense ground cover. At present its entire breeding range is confined to five or six counties in Michigan.

A male Kirtland's Warbler lands on the author's hand as a nest is examined. This picture, taken in 1950, would not be possible under strict regulations enforced today.

The total weight of all the Kirtland's Warblers in the world is less than twelve pounds. An adult weighs only a half ounce.

no nests, have been reported from Wisconsin in 1978, 1979, and 1980; from Ontario, in 1977 and 1978; and from Quebec, in 1978. Old migration records strongly suggest that in the nineteenth century the species may have had a larger breeding range, including suitable jack pine areas in Wisconsin and Ontario.

The first count, in 1951, indicated 432 males. In 1961, this count reached 502 males. Ten years later the total was 201. Starting that year, counts have been made annually. Over a twelve-year span, 1971 through 1982, the population of singing male Kirtland's has remained remarkably stable. The lowest count was in 1974, when only 167 males were reported, and the highest was 242 in 1980. The official count for 1983 placed the number of singing males in Michigan at 215. This was an increase of 8 over the 207 recorded in 1982.

Census taking consists of systematically searching for singing males in suitable habitat in mid-June. According to Harold Mayfield, author of a book devoted entirely to the species, this is a feasible method of estimating total population: count singing males and double the number, figuring one female for each male. As he cautions, however, this method may be off as much as 20 percent. Unmated females and males in the breeding area are imponderables.

Unquestionably the exacting demands of nesting Kirtland's Warblers have been responsible for limiting its population. In young jack pine forests (*Pinus banksiana*) Kirtland's Warbler will eventually survive or perish, for it is only in this very specialized environment that the warblers will nest. This drama has so caught the public's attention that Mayfield declares that it is one of the best understood songbirds in America.

The Jack Pine Warbler, as it is popularly known, has these special requirements for its nesting area: jack pines must predominate; the pines must be six to twenty feet tall; the pines must be in large stands, ideally two hundred acres or more; and the ground beneath and between the pines must be sandy and porous with low ground cover. The natural provider of these ideal conditions is forest fire. Since natural fires are stopped if possible, forest management practices for preserving the Kirtland's Warbler now include controlled burns.

The warbler population faces two other major problems in its struggle for survival: the Brown-headed Cowbird and a practically unknown factor, the dangers in its widespread winter range in the

The Brown-headed Cowbird problem that plagued Kirtland's Warblers for many years has been solved by using traps like this to capture the parasites before they lay eggs in warblers' nests.

Bahamas, an area of about forty-five hundred square miles in hundreds of scattered islands and cays. Although the wintering area has been searched on numerous occasions, investigators have had difficulty even finding the tiny bird in this sprawling range.

The Brown-headed Cowbird problem has been attacked successfully. As the cowbird extended its range northward in the last century, it found conditions ideal for parasitizing the Kirtland's Warbler. When a crisis was recognized in 1971, concerned friends of the warbler moved the next year to eliminate cowbirds by trapping them. That year, twenty-two hundred were taken in large chicken-wire traps and were asphyxiated. Since that time cowbirds have been destroyed each spring at Kirtland's nesting sites, and beneficial results appeared immediately in higher production of young warblers. However, of the hundreds of adults and young that leave Michigan in August and September, only about two hundred pairs have returned to breed the following year. The survival rate through two migrations and a winter spent in the tropics is about 65 percent for adults but unusually low for yearlings, and thus perplexing.

Female Kirtland's Warblers build their nests on the ground in a thick cover such as grass, sweet fern, and blueberries under jack pine trees.

This fledgling Kirtland's Warbler will spend the winter in the Bahamas with others of its kind. The survival rate for yearlings is low.

To the surprise of many not well acquainted with this bird, it does *not* nest *in* the jack pine trees that are so crucial to its breeding. The birds nest on the ground under the trees where the nest is hidden in a thick cover of grass, sweet fern, or blueberries. The female Kirtland's alone incubates the eggs for fourteen days, the longest incubation period reported for any North American warbler. The male feeds the incubating female and assists in feeding the young. It is known that occasionally pairs have had two successful broods in a summer, but this is not common.

Mayfield declares that Kirtland's Warbler "can scarcely be called an accomplished singer." He continues, "The song is not truly musical but, rather, loud, clear, emphatic, and frequently repeated. It has none of the buzz and trill so common among Wood Warblers, but reminds the listener of the chattering quality of a House Wren's song, though it is briefer. Field students are reminded of the Northern Waterthrush and some notes of the House Wren, but the resemblance usually is not close enough to cause one to mistake the Kirtland's Warbler for either."[1]

The first specimen of Kirtland's Warbler was given to Dr. Jared P. Kirtland by his son-in-law, Charles Pease, who had collected it near Cleveland, Ohio, on May 13, 1851. Kirtland turned the specimen over to Spencer F. Baird, who, in 1852, described the bird for science and gave it its name. Years later it was learned that Dr. Samuel Cabot of Boston had captured a male on shipboard near the Bahamas in October, 1841. The first nest and eggs were collected by James A. Parmalee on June 6, 1904, after which oologists paid as much as twenty-five dollars each for Kirtland's Warbler eggs. A nest with young birds had been discovered in 1903 by Norman A. Wood, but nests with eggs had more import to egg collectors of that period.

Breeding Range of Kirtland's Warbler

[1] Mayfield, Harold, *The Kirtland's Warbler* (Bloomfield Hills, Michigan: Cranbrook Institute of Science, 1960), p. 125.

28. Prairie Warbler

Dendroica discolor

PLATE 12

The depth of study that may be given a single species is indicated in a 595-page monograph by Val Nolan, Jr., entitled *The Ecology and Behavior of the Prairie Warbler* (*Dendroica discolor*). Nolan, a biology professor at Indiana University, pursued his study over fourteen years of intense field work. The results may be more than the average bird watcher wants to know about this species, but I have had no question about the Prairie Warbler that Nolan has failed to answer.

Judging by the habitat the bird prefers, *Prairie* seems to be a misnomer. It is not an inhabitant of western prairies or grassy plains, but apparently that is not what Alexander Wilson had in mind when he proposed its English name. Wilson found the species near Bowling Green in the "barrens of southwestern Kentucky," an area known to local residents as "prairie country."

The nesting environment preferred by the Prairie Warbler includes forest edges, dry, brushy clearings, pine barrens, sproutlands,

The author prepares to photograph the nest of a Prairie Warbler in silky dogwood in western Pennsylvania.

The Prairie Warbler is not an inhabitant of western prairies. It prefers forest edges, dry, brushy clearings, pine barrens (above), burned-over areas, and sproutlands.

burned-over areas, overgrown sand dunes, and Christmas tree plantings. In western Pennsylvania, I have observed that northward expansion of the Prairie's breeding range seems to be by extension from one Christmas tree planting to the next.

Migrants from the West Indies, Central America, and Florida are en route north by early March. One race of the Prairie Warbler is a year-round resident in mangroves bordering coastal marshes of southern Florida.

Prairie males have a large repertoire of songs, although two basic types are recognized. Nolan describes the first as "a series of short, equally spaced, rising, abrasive, jerky notes, typically about 15 in number, termed a chatter."[1] The second type is mostly two-parted, "a series of long notes and a series of short notes." Lynds Jones wrote

[1] Nolan, Val, Jr., *The Ecology and Behavior of the Prairie Warbler* (*Dendroica discolor*) (Washington, D.C.: Ornithological Monographs no. 26, American Ornithologists' Union, 1978), p. 67.

Prairie Warbler nests are usually built in forks of bushes and briers or on tree limbs or crotches, typically one to ten feet above ground. Gray-stemmed dogwood and honeysuckle were chosen here.

that "the notes seem to suggest *zee* syllables repeated six or seven times, deliberate at first, increasing to rapid at the close."[2] This last description fits the songs we used to hear when Prairie Warblers came tail-bobbing through the mangroves along the edge of our yard on Sanibel Island, Florida.

This tail-bobbing is not as vigorous as in Palm Warblers nor as pronounced as in waterthrushes, but it is an aid in identification.

Nolan has observed a habit of the Prairie that is uncommon among Wood Warblers: many build one or more "fragments" before building a complete nest. These fragments vary from a few shreds to well-made structures lacking only lining. He writes: "All but about 10 of approximately 125 fragments that I found looked indistinguishable from true nests on which a corresponding amount of work had been done."[3]

[2]Jones, Lynds, "Warbler Songs," *Wilson Bulletin,* vol. 12, no. 1, pp. 1–57, 1900.

[3]Nolan, p. 383.

Nests are usually placed in forks of branches in bushes and briers or on tree limbs or crotches, typically one to ten feet above ground, occasionally higher. The nest is a compact cup of plant down and bark shreds interwoven with fine grasses, held together with spider silk, and fastened to the supporting vegetation with webs and plant fiber. The female builds the entire nest, but the male has been seen inspecting the site and even making nest-shaping movements. Surprisingly, Nolan has seen female Prairie Warblers sing on as many as eleven different occasions.

Availability plays a part in the selection of nest material. A nest in Greene County, Georgia, near a cotton field, was built almost entirely of cotton. At Cape Hatteras, North Carolina, a nest was found that consisted largely of sheep's wool gathered from fragments snagged on nearby bushes. C. S. Brimley wrote that in North Carolina Prairie Warblers often use the gray leaves of rabbit tobacco (*Gnaphalium*) in constructing nests.

When predation occurs, female Prairie Warblers have been known to build as many as five different nests like this in a season in the effort to bring off a brood of young.

Breeding Range of Prairie Warbler

In 188 nests examined by Nolan during egg laying, 14 contained five eggs, 139 held four, and 35 had three. For forty-three complete clutches, Nolan determined the shortest incubation period to be 10.5 days and the longest 14.5 days, the mean being 12 days. Males assist in feeding young but are not known to incubate or brood. Most young fledge in 9 or 10 days. At least in the middle and southern portions of the range, females are often double brooded, provided their first nests escape predation. However, Nolan points out that nest predation is so extremely heavy in some areas that most Prairie females never have a chance to try a second brood. When predation occurs (most often by snakes) the female in most cases almost immediately builds a new nest and tries again. Many females build as many as five replacement nests a year. In an experimental situation, one female was known to build nine nests.

Like some other Wood Warblers, this species has been termed "colonial." This is the result of overcrowding in a highly desirable habitat rather than an instinct for gregariousness. Nests are always widely scattered even in ideal situations, and each male vigorously patrols his own territory.

The Prairie is a regular victim of the Brown-headed Cowbird. In a Michigan study, Lawrence H. Walkinshaw found eighteen nests, of which five were parasitized. Nolan's study of 336 nests included 80 that held single cowbird eggs and 12 with two each.

Polygyny is known among other species of Wood Warblers, but it may be more prevalent in Prairie Warblers than in any others. In 135 territories under study, Nolan found that as many as 10 percent of the males were polygynous at one time during the breeding season. Distance between nesting females of polygynous males averaged about 330 feet. Only twice did a female build within 160 feet of a nest of her mate's other female.

The Latin species name, *discolor,* means "of different colors." The type specimen was described by Louis Jean Pierre Vielliot in 1807. It was a wintering bird from the West Indies.

29. *Palm Warbler*

Dendroica palmarum

PLATE 12

It has been my good fortune to study and photograph Palm Warblers in a secluded bog on Mount Desert Island, Maine, locally called Great Heath (pronounced HAith by the natives). It is an extensive stretch of wet and spongy sphagnum moss and reindeer lichens embellished with sundew, pitcher plants, Labrador tea, bog rosemary, and stunted black spruce; and, in season, home of such lovely orchids as *Arethusa* and *Calopogon* (grass pink).

Over thirty years ago, James Bond, of the Philadelphia Academy of Natural Sciences, introduced me to this botanical wonderland, summer home of Palm Warblers, Lincoln's Sparrows, Common Yellowthroats, Hermit Thrushes, Yellow-bellied Flycatchers, and Black Ducks. Bond was expert at finding nests of Palm Warblers set deeply into hummocks of sphagnum moss, protected overhead by black spruces. In that boggy wilderness where every direction looked the same, Bond's only means of returning to a nest he had found was to snap the twig of a nearby tree or shrub. No other mark ever guided him.

Two subspecies of the Palm Warbler are recognized, and each is normally identifiable in the field: the Yellow Palm Warbler (*D. p. hypochrysea*) and the Western Palm Warbler (*D. p. palmarum*). The subspecies that breeds in Maine is the Yellow Palm, and since its fall migration takes it south along the Atlantic coast, it naturally is the subspecies one would expect to find in Florida, where flocks of Palms are common in winter. That is not the case. For some unexplainable reason, the Yellow Palms veer westward as they reach northern Florida and spend the winter along the Gulf Coast. Just as puzzling is the

James Bond, eminent Philadelphia ornithologist, introduced the author to the Palm Warbler in this boggy heath (*hayth*) on Mount Desert Island, Maine.

fact that the western form crosses the United States and spends the winter in Florida, Cuba, and the West Indies. Next to the Yellow-rumped, the Palm is the commonest warbler in southern Florida throughout the winter. From 3,500 to 5,400 individuals are recorded annually on Audubon Christmas counts in that state.

The Palm Warbler is a hardy bird. It moves from the warm comfort of the sunny southland to its breeding grounds in the northern United States and Canada long before the last winter blasts have ended. It arrives in New England in early April and nesting is under way in May. During that month, the temperature is near freezing on many mornings "down east" in Maine, and you can feel solid ice underfoot in the heath. My earliest records for Great Heath indicate that one female started to build her nest on May 5; four eggs hatched on May 30. In another instance, I found four young three days old on May 30, and another with five eggs hatching on May 29. Since I have records of Palm Warblers building nests as late as mid-June in Great Heath, I would not be surprised that this species is, at least occasionally, double brooded.

Some observers claim that the Palm is a colonial species. Nests I have found have always been far apart; thus I would describe the nesting as in "scattered pairs" in ideal habitat such as Great Heath. This appears to be true of the western race as well.

Palm Warblers are attracted to two types of nesting habitat. One is the sphagnum and spruce bogs typified by the Mount Desert location described here. The other, to which the western race seems more attracted than the Yellow Palm, is on dry plains of pines with clearings of low ground cover of blueberry, bearberry, sweet fern, and similar plants. In such a habitat, the Palm will occasionally build its nest in the low branches of small conifers. Even in a sphagnum bog, the birds sometimes place their nests in the lower limbs of black spruces, although ground nests are much more typical.

The song of the Palm is another warbler voice that is commonly compared to the song of the Chipping Sparrow. It is a trill, a series of short notes repeated rapidly with only a slight change in pitch. At the beginning of the nesting season, males have been watched singing

In a sphagnum bog, Palm Warbler nests are sunk deep in moss and protected overhead by stunted black spruces. Cowbirds rarely victimize them.

Early in the season the male Palm Warbler sings from high perches. His song is a series of short notes repeated rapidly with only a slight change in pitch.

from high perches. Later in the season, most songs are delivered from low perches, stunted spruces, or bushes.

Daniel A. Welsh studied the territories of ten Palm Warblers in a coastal bog in Nova Scotia, where he discovered one male that was polygynous. This individual fed the young of both females to which he was mated.

Friedmann considers the Palm Warbler a rare victim of the Brown-headed Cowbird. He cites only seven instances.

The Palm Warbler was discovered on the island of Santo Domingo, and early ornithologists thought it was a permanent resident of that and other islands of the West Indies. This undoubtedly accounts for the inappropriate English name of the bird. It was many years after it was described in 1788 that the Palm Warbler was recognized as a nesting species in the United States and Canada. Formerly, English names for this species were Redpoll Warbler and Little Yellow Wagtail—the former because of the bird's chestnut cap, which is so bright in spring; the latter for its characteristic tail-bobbing habit.

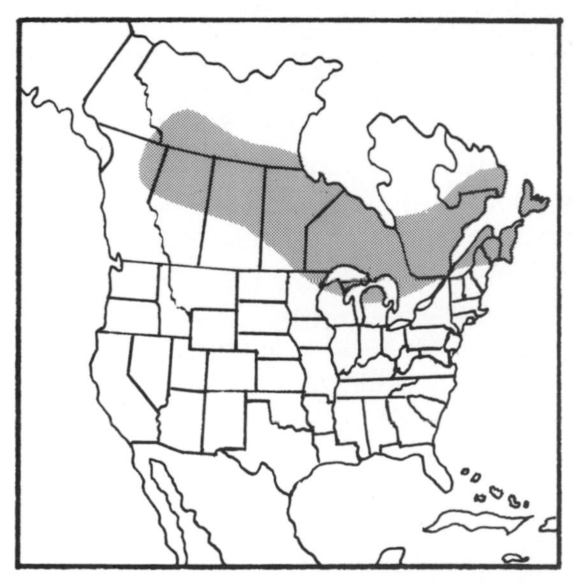

Breeding Range of Palm Warbler

30. Bay-breasted Warbler

Dendroica castanea

PLATE 15

At 10:30 A.M. on June 7 I discovered a female Bay-breasted Warbler sitting on a nest in a white spruce tree. I climbed fifteen feet to the horizontal limb it rested on and counted four eggs. I was disappointed because the Bay-breasted is one of three Maine Wood Warblers that frequently lay five, six, or even seven eggs. I needed the nest picture for my Peterson Field Guide, so I decided to photograph it anyway.

I walked to my car and returned with my camera and a ladder that would reach the nest limb easily. The female was on the nest again, apparently incubating. She left when I touched her back. It was now 10:50, and in those twenty minutes the Bay-breasted had laid her fifth egg. Rarely do I find warblers so accommodating.

The Bay-breasted Warbler's output of eggs, and thus of nestlings, like the Cape May Warbler's, seems to be correlated to the abundance or absence of the spruce budworm, at least in areas where both the warblers and budworms are usually in great abundance, such as northern New England and southeastern Canada. Population densities of Bay-breasteds and Cape Mays increase during infestations of the spruce budworm, the larva of a tortricid moth, a destructive pest in many northern coniferous forests. Conversely, in years when there are no budworm outbreaks, the numbers of these warblers may dwindle or they may even disappear in areas where previously they were numerous. Since Bay-breasted Warbler populations are known to fluctuate in noninfested areas, it is not known definitely how much impact the budworm phenomenon has on the up and down cycles of this species.

When I found this nest in a white spruce it held four eggs. Twenty minutes later there were five. Bay-breasted Warblers may lay six or, rarely, seven eggs in a clutch.

Researchers have determined that the astronomical number of spruce budworms available as food for birds during an outbreak far exceeds the demands of all the birds that inhabit the infested areas. So why don't more birds nest there? Why not take advantage of this continuous and abundant food supply for themselves and their young? Douglass H. Morse suggests three reasons: 1. numbers of birds are inadequate to populate the areas more densely, 2. budworms do not provide a complete diet for the birds, and 3. territorial behavior may be limiting numbers.[1]

The Bay-breasted, one of the largest Wood Warblers, is a late migrant from its winter home in central and eastern Panama, northern Colombia, and western Venezuela. Unlike several other boreal warblers the Bay-breasted does not extend its range southward along the Appalachian Mountains. It prefers thick stands of spruce and fir, but

[1] Morse, Douglass H., "Populations of Bay-breasted and Cape May Warblers During an Outbreak of the Spruce Budworm," *Wilson Bulletin,* vol. 90, no. 3, p. 411, 1978.

A Bay-breasted Warbler brings a mouthful of spruce budworms to nestlings in a coniferous forest. These larvae often are the main source of food for this warbler.

The female Bay-breasted Warbler saddles her nest to a horizontal limb fifteen to twenty-five feet above ground in a dense conifer. The outer walls are loosely woven of conifer twigs.

where spruces are not plentiful, the birds accept mixed forests of birches, maples, firs, and pines.

This species commonly seeks the middle branches of spruces for both foraging and nesting. Its movements while foraging have been described as slow and sluggish, and to follow a male through his morning exercises is to realize that he moves very little in contrast to other warblers in the same area.

The Bay-breasted Warbler's nest is large, with outer walls loosely woven of conifer twigs that protrude on all sides. It is saddled on a horizontal limb in a dense conifer and is typically from fifteen to twenty-five feet above the ground. Of eight nests that I have found in Maine, the highest was thirty-five feet; the lowest was nine feet; the average was nineteen feet above the ground.

At one nest where I photographed the adults feeding nestlings, the four youngsters fledged on the tenth day after hatching. During the last day they were in the nest, they were restless and moved often, preening, fluffing feathers, spreading their wings; and on numerous occasions, the birds snapped at mosquitoes above the nest. Once

During the summer, the Bay-breasted Warbler prefers dense stands of spruce and fir. Where these are not available, they accept mixed forests of birches, maples, firs, and pines.

when the female approached the nest, she suddenly froze in one position. With spruce budworms dangling from her bill, the bird remained absolutely motionless for about fifteen minutes. The reason, I discovered, was a red squirrel in the nest tree.

I must confess that I cannot distinguish between the songs of the Bay-breasted and Cape May Warblers. It seems to me that the former is a bit louder, but both birds seem to lisp the phrase *zee* or *chee.* Sometimes I think it says a high-pitched, wiry *cheat it, cheat it.* My wife does not share my problem. To her, the song is definitely *tisit tisit tisit tisit.* Some birds accent the *tis* and others the *sit.* Unlike most female warblers, the female Bay-breasted sometimes sings, usually from the nest and mostly in answer to the male's song.

Three hybrids involving the Bay-breasted Warbler have been reported: two crosses between this species and the closely related Blackpoll Warbler and a cross with the Myrtle Warbler.

Both Wilson and Audubon considered the bird rare. In his *American Ornithology* (1832) Alexander Wilson called the species "Bartram's Little Chocolate-breasted Titmouse." Its present species name seems more apt. *Castanea* is Latin for "chestnut," alluding to the color of the bird's breeding plumage.

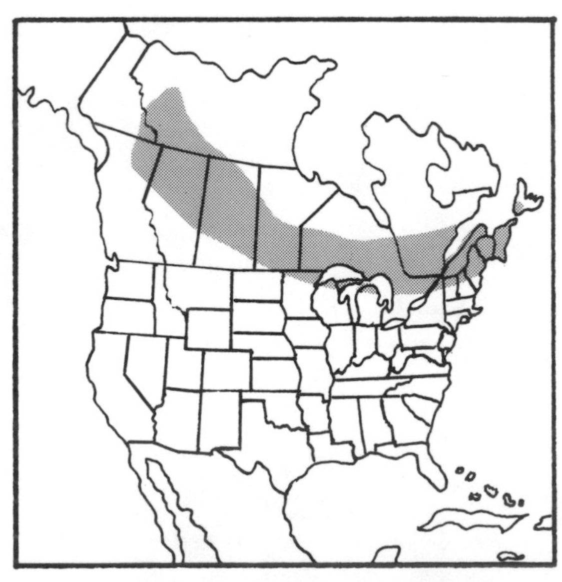

Breeding Range of Bay-breasted Warbler

31. Blackpoll Warbler

Dendroica striata

PLATE 14

In my life with birds, two events concerning Blackpoll Warblers are memorable. One was good; the other miserable.

First, the good one. In July 1950, the Long family from Maine and the Harrison family from Pennsylvania pitched tents for a week on Little Duck Island, seven miles off the coast of Mount Desert Island, Maine. Little Duck, which today is owned and protected by the National Audubon Society, is a nesting area for great numbers of Leach's Storm-Petrels, Common Eiders, Herring and Great Black-backed Gulls, Black Guillemots, Double-crested Cormorants, and others, including a few Wood Warblers. Our purpose in camping there was to photograph resident birds.

The outstanding event of this expedition was the finding of a Blackpoll Warbler's nest containing young birds. The nest was thirty-three feet above ground in a white spruce; as far as I know, this is the highest Blackpoll nest ever recorded. The nest held another record at the time: the southernmost nesting for coastal populations of Blackpolls. Since then, Ralph Long has found a nest at Wonderland on Mount Desert Island.

The miserable event happened in 1968 when Long and I traveled to Rangeley Lake, Maine, to spend two days finding and photographing the nest and eggs of a Blackpoll Warbler. My companion had heard Blackpolls singing in that area a week earlier, so we anticipated little trouble in locating a nest. What we did not anticipate was rain, torrents and torrents of it. For breeding, Blackpoll Warblers are attracted to low coniferous forests, especially spruces. Their nests are normally placed low in young conifers, commonly two to seven feet

The Blackpoll Warbler's nest is usually against the trunk of the chosen tree, supported by horizontal branches, and concealed by overhanging foliage.

Nest and eggs of the Blackpoll Warbler, for which we searched without success during two days of rain near Rangeley, Maine.

Typical Blackpoll Warbler nesting habitat. Nests are placed low in young conifers like these in the understory.

above ground against the trunk of the tree, supported by horizontal branches and well concealed by overhanging foliage. We had acres of young spruces through which to search, and every tree was saturated with water as only a spruce tree can be. In order to see a nest we were forced to push in through water-logged branches to the trunk, and we were soaked. The water came from the sky above and from the trees that surrounded us. Had we been looking for Magnolia Warbler nests, which are commonly out on the branches, we would have been wet, but not drenched. Never have I been so wet, cold, and miserable. We stuck it out for two days, drying our clothes as best we could at night and putting them on, ready or not, in the morning. Blackpolls? Yes, we heard a few males singing. Nests? None.

This fledgling Blackpoll Warbler faces a southern migration that may carry it twenty-five hundred miles south of the nest where it was born.

The male Blackpoll's common song is weak, high-pitched, and monotonous—a series of notes all on one pitch, loud in the middle, soft at the end.

The diminutive Blackpoll Warbler has an international reputation for being the "travelingest" of all land birds, a reputation rivaled only by the Arctic Tern, a water bird, and the Golden Plover, a shore bird. No other warbler, and not many birds of any kind, fly as far in migration. It has been estimated that no Blackpoll has a migration route of less than twenty-five hundred miles north and twenty-five hundred back again, for Blackpolls winter in South America and breed as far north as Alaska and the Arctic Zone in Canada. Those that travel from Alaska to Brazil undoubtedly fly over ten thousand miles round-trip each year.

Spring migration brings hordes of Blackpoll Warblers from South America, across the West Indies and into Florida. From Florida they move northward over a broad front to their territories in the spruce forests of the northern United States and Canada. By August the birds have left Alaska and the far north, but when they reach the United States migration slows and Blackpoll Warblers are seen throughout the autumn months in many of the mid-Atlantic and southern states. Even birds that nest west of the Mississippi Valley typically move toward the Atlantic Coast en route south.

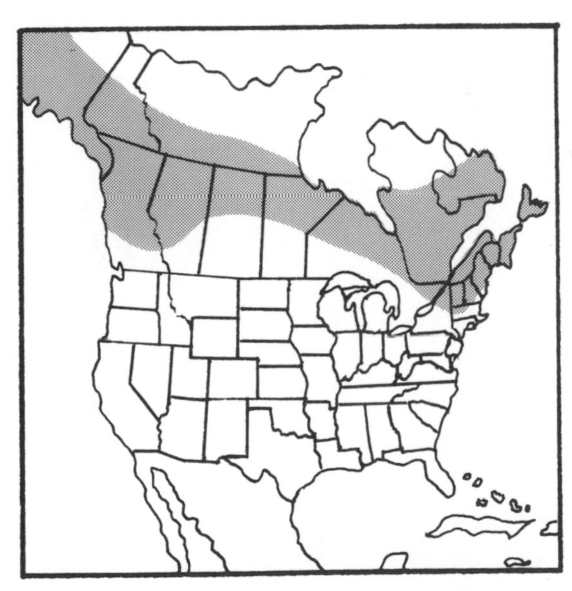

Breeding Range of Blackpoll Warbler

The male Blackpoll's song is weak, high-pitched, and monotonous. The common version is a series of notes all on the same pitch, growing louder in the middle, and then softer to the end. In another song, the notes are a rapid trill. Albert Rich Brand measured the vibration frequency of songs of fifty-nine passerine birds and found the song of the Blackpoll to have the highest pitch of any he studied. It is well beyond the range of hearing of many people.

Alexander Wilson described a species which he named Autumn Warbler, *Sylvia autumnalis,* which has since been identified as a Blackpoll Warbler in fall plumage. The immature and winter plumages of the Blackpoll and Bay-breasted Warblers are not readily distinguishable, although the males are very different in spring. The species name, *striata* (stry-AH-tah), is Latin for "striped," very suitable for a bird with stripes on its back, rump, and flanks.

32. *Cerulean Warbler*

Dendroica cerulea

PLATE 17

Finding the nest of a Cerulean Warbler is difficult enough, but photographing the adults at their nest or even looking inside a nest to determine its contents is a task for the adventurous and the brave. No pair of Ceruleans that I have encountered has made the task easy. All twelve nests that I have found in Pennsylvania have been hidden high in a deciduous tree and far from the trunk on a horizontal limb. All except one were over forty feet above the ground (the highest was sixty feet) and all were from six to twenty feet from the trunk.

Except for one in a sugar maple and one in a shagbark hickory, all my nests were in oaks. All were in open woodlands with little or no ground cover beneath the trees. In other parts of the Cerulean Warbler's summer range, elms, black ash, beech, basswood, or sycamores may be the predominant trees, but whatever the tree, the bird's outstanding natural tendency for nesting high remains unaltered.

Only two of my nests were accessible, and both contained cowbird eggs. My sampling was too small to be compared with Friedmann's record of eighteen parasitized nests.

The Cerulean Warbler's nest is dainty and compactly built, a neat knotlike structure that resembles nests of an Eastern Wood-Pewee or a Blue-gray Gnatcatcher. This shallow nest made of mosses and lichens interwoven with spider silk is most unwarblerlike in appearance.

What part does a male play in the female's selection of the site for her nest? Most students have decided that the answer is very little, if any. An observation of a Cerulean Warbler male by G. B. Van Cleve of Pittsburgh is interesting. He reported seeing the male spend some time investigating the junction of a nearly horizontal forked branch of

Cerulean Warblers often nest in oak trees in open woodlands with little or no ground cover beneath the trees. Most nests are forty to sixty feet above ground and six to twenty feet from the trunk.

an oak, fifty feet above ground. The male assumed a squatting position and began turning about in an irregular fashion which suggested that he was forming a nest. A female suddenly dashed to him whereupon the male immediately departed leaving his mate (presumed) inspecting the situation. She turned about on the same spot and then left. Van Cleve asked, "Was the male suggesting a nesting location; or was this sort of pantomime courtship display?"[1]

From its winter home in South America, principally Colombia, Venezuela, Bolivia, and Peru, the Cerulean Warbler moves northward to enter the United States in Texas and Louisiana in April, reaching its nesting grounds in May.

[1] Personal correspondence.

Nests of the Cerulean Warbler are compactly built, knotlike structures that resemble nests of an Eastern Wood-Pewee or Blue-gray Gnatcatcher.

The male Cerulean Warbler has little to do with site selection and nest building, but he takes part in feeding nestlings.

On his territory, the male is an incessant singer from dawn until dusk. This is a fortunate circumstance for those who want to see him, for he is hard to find high in the treetops where he spends most of his time. The song has been compared to that of the Northern Parula. It consists of three or four notes all on one key, followed by one that is longer and higher in pitch: *zray zray zray zray zray zreeeee.* I would compare it to the song of the Black-throated Blue Warbler.

The Cerulean Warbler was named and described by Alexander Wilson in *American Ornithology.* He described only the male. The specimen had been sent to him by Charles Wilson Peale, of the family of illustrious painters, who had collected it in eastern Pennsylvania. The female was described later by Charles Lucien Bonaparte. The species name *cerulea* (see-RULE-ee-ah) is Latin for "blue" (like the sky), an appropriate name, for no other North American Wood Warbler has a similar cerulean-blue back.

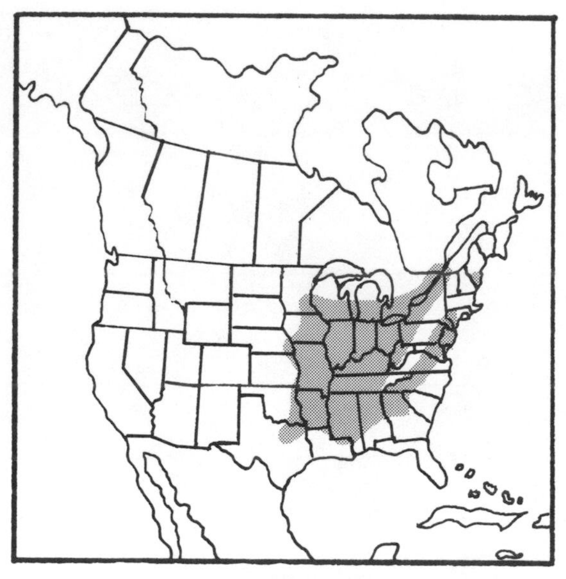

Breeding Range of Cerulean Warbler

33. *Black-and-white Warbler*

Mniotilta varia

PLATE 20

Winter-weary folks who search for every event that forecasts the return of warm, pleasant weather greet the Black-and-white Warbler with enthusiasm. Along with the Louisiana Waterthrush, it is among the first of the Wood Warblers to arrive at its nesting grounds in the North each year and is a welcome herald of spring. The return journey of the Black-and-white may be relatively short, because its winter home could have been as near as the Carolinas or south to Florida. However, it may have traveled from the West Indies, Central America, or even northern South America.

According to Robert C. Leberman, who has kept records for many years, the Black-and-white Warbler reaches western Pennsylvania during the third week in April, the nineteenth on the average. There is a practical reason why this warbler can safely return long before most Wood Warblers are capable of facing the harsh conditions of early spring. While most warblers depend on insects and larvae that appear with the opening of leaves, the Black-and-white feeds on wood-boring insects, bark beetles, moths, and many dormant insect larvae, which it can find on trunks and limbs of bare trees. Its behavior is reminiscent of the creeping habits of the nuthatch, Brown Creeper, and Pine Warbler. This habit gained it a now obsolete English name, Black-and-white Creeper. The only other bird in its breeding range that might be mistaken for the Black-and-white is the Blackpoll Warbler; but the Blackpole, like the chickadee, has a solid black cap, and its behavior is different.

The Black-and-white Warbler is one of the first Wood Warblers to arrive in the North in spring. In Pennsylvania, the third week in April is the average arrival date.

In addition to its unwarblerlike creeping, the Black-and-white has a less conspicuous attribute that is not found in many other warblers. It has rictal bristles, long bristles about its mouth that apparently enlarge its gape. After leaves begin to open and more flying insects appear, the Black-and-white adds fly catching to its food-garnering behavior and these bristles help capture flying insects. One theory has been that the bristles may serve as a funnel directing insects to the mouth. Another suggests that rictal bristles are highly sensitive to touch and may alert the bird to the presence of food even if the bird has not seen it. The Canada Warbler and the American Redstart, among others, also possess these long bristles.

The female Black-and-white Warbler alone builds the nest. She incubates the four or five eggs for eleven or twelve days. One brood a year is normal.

Upon arriving on its nesting territory, the male Black-and-white Warbler sings his wheezy, wiry song incessantly until nesting is well under way. His thin, high-pitched voice may be represented by the words *wee-see, wee-see, wee-see, wee-see, wee-see.* Albert Rich Brand, judging from his mechanically recorded songs of warblers, placed the Black-and-white's song fourth highest in pitch among the sixteen he recorded. The Blackpoll, Blue-winged and Blackburnian are higher. Winsor Marrett Tyler wrote: "Its song is made up of a series of squeaky couplets given with a back-and-forth rhythm, a seesawing effect, like the Ovenbird's song played on a fine delicate instrument."[1] In June, when nesting is well along, the Black-and-white becomes very quiet and may be overlooked.

The Black-and-white female alone builds the nest; it is on the ground, typically at the base of a tree, stump, or large rock, under a log, or under a fallen tree branch. It is concealed in a drift of leaves

[1] Tyler, Winsor Marrett, "Black-and-white Warbler," in Bent, p. 10.

When young leave the nest, they are recognizable as Black-and-white Warblers, although they are slightly tinged with brown.

and usually hidden from above. A few scattered records tell of nests that have been built on stumps as high as fifteen inches above ground.

When disturbed at her nest, the female flutters across the ground displaying a typical "broken wing act" (described in chapter 38). Later, when the adults are feeding nestlings, it has been my experience that they are not shy about approaching the nest in the presence of an onlooker.

When the young leave the nest, they are clearly recognizable as Black-and-white Warblers although they are slightly tinged with brown. Even the head stripes are evident.

It is unusual that a Black-and-white Warbler was observed feeding young in the nest of a Worm-eating Warbler in Hocking County, Ohio. The Worm-eating Warblers resented the intruder and attacked it when it came to the nest. Attacks failed to stop the Black-and-white from feeding. In another instance, S. Charles Kendeigh watched a pair of Black-and-whites feed a young Ovenbird that was out of the nest and apparently separated from its parents. (It may simply be coincidence, but it is interesting to note that young Worm-eating Warblers, Ovenbirds, and Black-and-white warblers all have dingy head stripes.)

The Brown-headed Cowbird commonly lays eggs in Black-and-white Warblers' nests. A record number of eight cowbird eggs and two warbler eggs was reported in a Michigan nest with the female incubating all ten. Another nest contained five cowbird and two warbler eggs.

In a list of longevity statistics on North American birds, John H. Kennard reported that among Wood Warblers, the Black-and-white Warbler holds the record. A female banded in Hillsboro, North Carolina, on September 1, 1957, was recovered near Philadelphia, Pennsylvania, on or about September 21, 1968. The estimated age of the bird was eleven years and three months. The average old age of the twenty-six species of Wood Warblers Kendeigh listed is approximately six years. The average age of *all* warblers would be considerably less than six years, for this study included only banded birds four years old or more.

This Black-and-white Warbler's nest with five warbler eggs escaped cowbird parasitism, although as many as eight cowbird eggs have been found in a nest. This site, under the protection of a fallen branch, is characteristic.

An odd-looking Wood Warbler collected in Louisiana in 1954 was thought at first to be an aberrant Black-and-white Warbler. Later, Dr. Kenneth C. Parkes of the Carnegie Museum, Pittsburgh, declared the specimen to be an intergeneric hybrid between a Black-and-white and a Cerulean Warbler.

The generic name *Mniotilta* (nih-oh-TILL-tah) comes from two Greek words, *mnion* meaning "moss" and *tillein* meaning "to pluck or pull out." The species name *varia* (VAY-rih-ah) is Greek for variegated. Loosely translated, it means black and white moss plucker. It is surprising that neither the English nor the scientific name refers to the well-known creeping habits of the species. It was one of the last species to be described by Linnaeus himself.

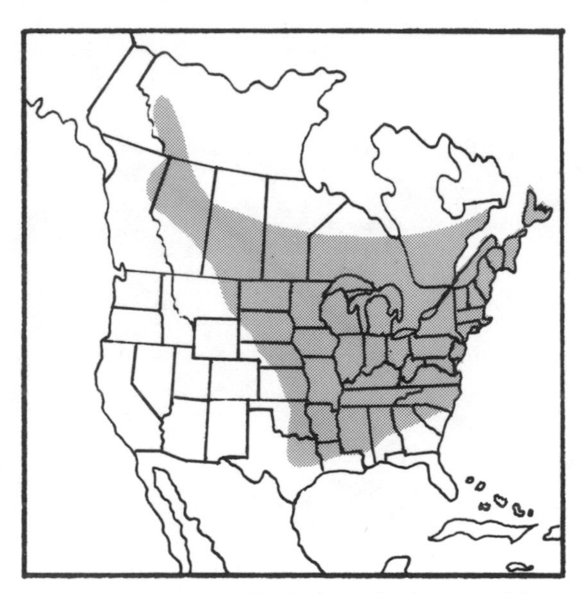

Breeding Range of Black-and-white Warbler

Nesting habitats for the Black-and-white Warbler include both deciduous and coniferous woodlands, especially hillsides and ravines.

34. *American Redstart*

Setophaga ruticilla

PLATE 16

Although it happened many years ago, I still remember vividly the first time I saw a redstart. It was late in May in a Pennsylvania woodlot of young maples. A flaming black and orange torch came flickering through the leaves on a shaft of sunlight. I first spied him high in a tree, darting out in flycatcher fashion to snare passing insects. One victim must have eluded him, for he fluttered down through the sun-spangled leaves, fanning his tail, spreading his wings, and displaying his brilliant plumage. Down he came like a tropical butterfly only to swoop upward suddenly and perch near his mate.

What a lovely pair of birds! Little wonder that the Cubans call the American Redstart *candelita,* "little candle," and that elsewhere in the tropics redstarts are known as *mariposas,* "butterflies." Catesby and Edwards both described the American Redstart, and in 1758 Linnaeus named it from their descriptions.

The name *redstart* comes from the fancied likeness of this bird to the European Redstart, which is a thrush and a far cry from our American Redstart. About all the two have in common is a habit of flicking their tails and exposing colorful outer feathers. The generic name, *Setophaga* (see-TOFF-ah-gah), comes from the Greek *ses,* insect, and *phagein,* to eat. The species name, *ruticilla* (root-ih-SILL-ah), indicates a red tail.

This bright color is a comparatively long time in coming to the redstart even though the young molt from nestling or juvenal plumage earlier than most warblers. George A. Petrides raised a nestling and found that by twenty-two days of age it was in a heavy molt from the grayish juvenal to greenish plumage. This molt had begun even before all the juvenal feathers had appeared.

What a lovely pair of birds! Cubans call the American Redstart *candelita,* "little torch." Others in the tropics call it *mariposa,* "butterfly."

Male American Redstarts do not acquire full adult plumage until after their first breeding season. A young male can be told from a female by black flecks on his throat and breast (tips of a few scattered black feathers) and tinges of orange that often border the yellow of the tail and sides. A few young males assume these colors in the fall of the year they are hatched, but most do not acquire them until a molt just before breeding when they are nearly a year old. Although these first-year males resemble females, they breed in this plumage; as a group, however, they are not as successful as the more aggressive adult males.

For nesting, American Redstarts like deciduous woodlands better than conifers, but they are found in both. They are attracted to open hardwood thickets, roadside trees, tall shrubbery, bushy and tree-lined stream banks and ponds, and willow and alder thickets. The

If this fledgling American Redstart is a male, it will not acquire adult plumage until after its first breeding season, a trait shared with a number of Wood Warblers.

For nesting, American Redstarts like deciduous woodlands better than conifers, but they are found in both. Nests are often placed in the fork of a roadside tree ten to twenty feet above ground.

great fire that threatened Bar Harbor, Maine, in 1947 destroyed hundreds of acres of conifers. The resurgent growth is made up largely of maples, birch, and aspen. These trees are now large enough to suit American Redstarts, and the local population has increased.

Redstarts use a variety of nesting sites, but in some areas certain trees and shrubs are more popular. I worked in an area in western Pennsylvania where grapevines climbed up and over much of the woodland vegetation. Seven of twelve American Redstart nests in the area were built in these vines. In Maine, I found twenty-one nests, of which nine were in birch, six in alder, three in spruce, and one each in tamarack, willow, and elm—a far cry from Kirtland's Warbler, which will nest only under a jack pine tree.

Millicent S. Ficken, who studied twelve pairs of American Redstarts during three seasons at Ithaca, New York, divided courtship into three periods: pair formation, pre-nest building, and nest building. Pair formation is accomplished unbelievably quickly, the female

American Redstart male at nest built in a grapevine. Note cotton used in nest walls.

The American Redstart is a common victim of the Brown-headed Cowbird, but this nest escaped. Three or four eggs is a normal clutch.

reaching the territory at night and having accepted a mate by the following morning. This period includes some aggressive behavior on the part of the male and rejection by the female. During pre-nest building the pair spends considerable time chasing, and the female acquaints herself with his territory before selecting a nest site. Nest building begins as soon as the female selects a site. Usually, when the nest is almost finished, the male performs various displays and copulation follows. Pairs seem to copulate only once or twice per season unless the nest is destroyed and they start a new one. All sexual activity ceases the day before the first egg is laid.

An American Redstart's nest is a felted cup placed from ten to twenty feet above ground in a fork of a tree or shrub. The Yellow Warbler builds a similar nest, but with thicker walls, especially at the rim. The nests of both these species are higher than they are wide. The American Goldfinch nest, also similar, is wider than it is high.

This species is a common victim of the Brown-headed Cowbird. Of thirty-four nests found by Friedmann at Ithaca, twenty-three contained one or more cowbird eggs. Like the Yellow Warbler, the American Redstart sometimes builds a new floor to its nest to cover a cowbird egg. Millicent S. Ficken tells me that a female American Redstart may desert her nest if a cowbird lays an egg in it after the female has laid one or two eggs.

J. Claire Wood offers an interesting account of a Michigan nest with seven eggs which was shared by two pairs of American Redstarts. Wood writes: "When found one female was upon the nest and the other perched close beside it. They were equally demonstrative of anxiety as I ascended the tree. The eggs were in two layers and slightly incubated. This was not a case of polygamy, as both males were present. All four were living in perfect harmony."[1]

Redstarts appear to be strictly insectivorous. They are flycatching warblers, but also glean industriously among leaves and along branches and trunks. As the bird feeds, it indulges in feats of agility that few other warblers can approach, darting out suddenly for a passing insect, or fluttering in mid-air to pluck a dangling caterpillar from its web.

The song of the American Redstart is somewhat like that of the Yellow Warbler but shorter and less robust. Sometimes it reminds me

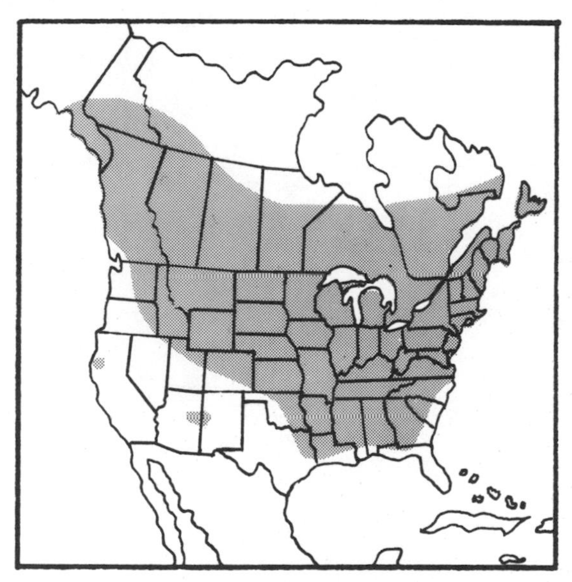

Breeding Range of American Redstart

[1] In Bent, ed., p. 665.

of the song of the Chestnut-sided Warbler; at other times it sounds like that of the Magnolia Warbler. One series of notes given all on one pitch reminds me of the Black-and-white Warbler. It can be utterly confusing. A clue to the singer's identity is that the American Redstart is the only Wood Warbler that habitually alternates between two different songs during the course of a singing session.

American Redstarts may leave their nesting range as early as August, but it is well into September before a general movement toward winter quarters is under way. The birds fly southward more or less directly over a broad front of twenty-five hundred miles, eventually leaving the United States from Florida or the Gulf Coast to spend the winter in Mexico, Cuba, Puerto Rico, Central America, and northern South America.

35. Prothonotary Warbler

Protonotaria citrea

PLATE 20

The story is told that when the Creoles discovered a brilliant golden warbler in the gray gloom above the dark waters of their Louisiana cypress swamps, they envisioned it as resembling the yellow robes worn by a *protonotarius,* an adviser to the pope. They named it *protonotaria.* Some time later, the English name became prothonotary. The species name, *citrea* (SIGH-tree-ah) is Latin for "pertaining to the citrus tree" and, loosely interpreted, to the lemon or yellow color of its fruit.

Historically a bird of southern swamps, flooded bottomlands, and borders of lakes, rivers, and ponds, the Prothonotary Warbler has extended its breeding range, especially in the Mississippi Valley and vicinity. It now nests as far north as Wisconsin and Minnesota. There as in the South it almost always chooses a swampy location.

A typical nesting site is a cavity in a rotted stub, often surrounded by water. It is the only eastern Wood Warbler that nests in cavities. (Lucy's Warbler, a western species, is the only other hole-nesting warbler.)

Not all Prothonotary Warblers are typical, and the list of unusual nesting sites used by this species is a long one. Samuel A. Grimes writes to me, "I have put up gourds for Prothonotary Warblers for many years and have had several pairs nest in them. Some have been in dense hardwood swamps and others in cypress 'bayheads.'" (A bayhead is a stand or island of bay trees surrounded by cypress swamp.) Here are some other sites this species has used: a small paper sack partly filled with staples left on a shelf near an open win-

Prothonotary Warblers may be attracted to man-made nesting sites such as birdboxes. Here a pair chose to nest in a hollowed gourd.

dow, a mailbox, a coffee can, a cheese box, an old enamel coffee pot hung in a gazebo, an old hornets' nest, a large glass jar on its side, the pocket of a pair of pants hanging on a clothesline, and a metal pail hanging on a porch.

One pair chose a ferry boat at Mammoth Cave National Park for their nest. The ferry was in operation continuously, moving from one side of the river to the other; but the birds continued to attend the nest, which was built in a guard over a pulley wheel.

In *The Oologist,* M. G. Vaiden told of a Prothonotary Warbler nest in a four- by ten-inch heavy metal toolbox mounted on one of the upright supports of a large log-loader in use along the Mississippi River. On May 19, 1936, when the nest was found, it held five eggs. A few days later the eggs hatched, and the parents fed the nestlings in spite of the jerking motion of the machinery and the nearness of the operator as he pulled logs from the river. All five young left the nest in due time. In 1937, Prothonotaries nested in the box again; but in 1938, a pair of Carolina Wrens built in the toolbox first, so the Prothonotaries used a small gasoline inlet pipe on a nearby unused houseboat.

Man-made nest boxes have proved very attractive when placed in well-chosen locations. Merit B. Skaggs has given me dimensions for boxes that he found successful in attracting the species along the Cuyahoga River in Geauga County, Ohio. The inside dimensions are 3.5 in. by 3.5 in. with a depth of 7 in. The entrance should have a diameter of 1.75 in. It does not seem to matter whether a perch is supplied. Ideally, the boxes should be on trees or stubs over water. In natural cavities, the birds commonly select holes from one to twelve feet above ground or water.

The Prothonotary's habit of nesting in holes has not made it entirely immune from Brown-headed Cowbird parasitization. Fried-

Typical nesting site for the Prothonotary Warbler is a cavity in a rotted stub, often surrounded by water. Nests have been found in surprisingly odd places, however.

A bird of southern swamps, flooded bottomlands, and borders of lakes, rivers, and ponds, the Prothonotary Warbler has extended its range north to Wisconsin and Minnesota.

mann reported no less than fifty-four definite records. In twenty-eight Prothonotary nests studied by Lawrence H. Walkinshaw, three were parasitized.

In this study made by Walkinshaw of twenty-eight nests along the Battle Creek River in Calhoun County, Michigan, in 1937, nineteen were in birdhouses above running water. The others included six in stubs over water and three in natural holes away from the riverbank. Walkinshaw reported that moss made up the bulk of material put in the cavity, the nest proper being on top of the moss. The nests were built almost entirely by females, but occasionally he observed males carrying moss to a site.

Walkinshaw tells me that in one birdhouse that he erected at Reelfoot Lake, Tennessee, three pairs of Prothonotaries nested in one season: one in May, another in June, and a third in July. Most of the pairs he observed in the South raised two broods. In Michigan, however, Walkinshaw reports that pairs rarely raised or even attempted two broods. Nest failures were common in Michigan, often because of House Wren interference.

Walkinshaw relates a surprising fact: young Prothonotaries are expert swimmers. If they happen to jump out of the nest cavity into the water prematurely, they are able to swim to shore or to a log or some object where they are safe.

The male Prothonotary Warbler is a constant singer throughout the nesting period. The song is distinct and loud, and to my ears says *sweet sweet sweet sweet sweet.* At a distance it sounds like a Spotted Sandpiper.

Migration southward begins as early as July, and by mid-August

The Prothonotary Warbler is one of two cavity-nesting Wood Warblers in the United States. The other is Lucy's Warbler, a desert species.

most Prothonotary Warblers are on their way to winter in Central America and northwestern South America. Eugene S. Morton points out two unusual characteristics of the Prothonotary as a wintering species in the tropics: First, pair bonds are evident, a situation unique among parulines; and second, Prothonotaries roost communally, even gathering in the same tree at dusk before flying to their final roost. "In this respect," Morton declares, "they are more like Orchard Orioles or other gregarious roosting icterids than warblers."[1]

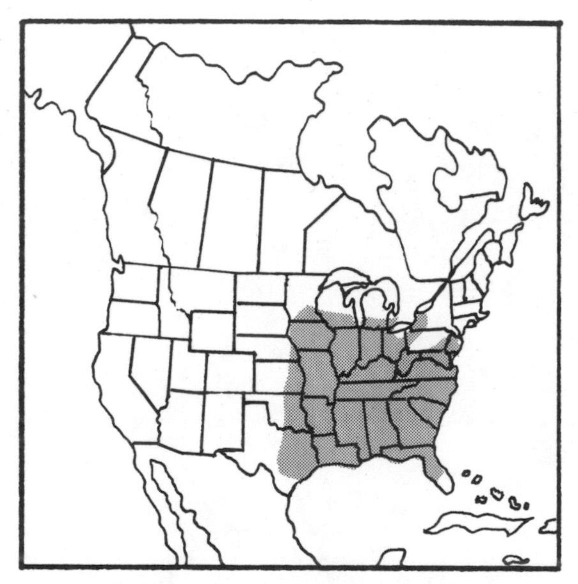

Breeding Range of Prothonotary Warbler

[1] Keast, Allen, and Eugene S. Morton, "Ecology, Behavior, Distribution and Conservation," in Keast and Morton, eds., p. 449.

36. *Worm-eating Warbler*

Helmitheros vermivorus

PLATE 18

In chapter 7, I stressed the fact that a Virginia's Warbler nest is one of the most difficult to find of all ground-nesting Wood Warblers. I believe I can safely put the nest of the Worm-eating Warbler in the same category. The general habitat of Virginia's (steep mountainsides) may offer more problems in searching for the nest of that species, but since Worm-eating Warblers invariably choose leafy hillsides for their nests, the difference may be slight.

I have searched long and hard for the nest of this bird and have found exactly two. The first was found on June 25, 1957, on a steep hillside in Butler County, Pennsylvania. I almost stepped on the female before she flushed at my feet. The nest held five eggs, which hatched on June 30. The second nest was found on May 29, 1982, on an almost-perpendicular hillside in North Park, Allegheny County, Pennsylvania. When it contained three eggs, it was destroyed by a predator.

During the twenty-six years that intervened between the finding of these two nests, I spent much effort and time in the search without success. At one time, I enlisted the aid of a dozen members of the Brooks Bird Club to help me comb the hillsides along the Ohio River near Wheeling, West Virginia. All were experienced birders, male Worm-eating Warblers were singing in the area, yet, no one found a nest.

I have never been happy with the English name of this species. Like most parulines, these warblers subsist mainly on caterpillars and other larvae, weevils, beetles, and bugs, which are loosely called worms by many people; but technically this is incorrect. The scientific

I have searched long and hard for the Worm-eating Warbler's nest. This is one of only two I have found.

Like most parulines, Worm-eating Warblers subsist on caterpillars, weevils, beetles, bugs, and adult and larval insects, loosely called worms.

name of the species is no more satisfying than the English name. *Helmitheros vermivorus* gives double emphasis to the misnomer. *Helmitheros* (hell-mih-THEE-roze) is Greek for worm-hunter; *vermivorus* (ver-MIV-oh-rus) is Latin for worm-eater; thus, a worm-hunting worm-eater.

The Worm-eating Warbler is an unobtrusive bird that must be sought to be seen. It is not particularly shy, but its olive green coloring above and buff below make it inconspicuous as it feeds on or near the ground in dense underbrush and thickets. Like the Ovenbird, which it resembles to some extent, the Worm-eating does not hop, it walks; and like the Ovenbird, it is mainly terrestrial.

The male's song is thin, rapid, and short, an unmusical buzz with an abrupt ending. It is most often likened to the more musical song of the Chipping Sparrow. A not infallible rule is: If you hear a Chipping Sparrow singing on a woody hillside, it's a Worm-eating Warbler. One observer noted that the sparrow has a rattle in its song rather than the buzz of the warbler. Fortunately, the two seldom occur in the same habitat.

Some observers have commented on the flight song of the Worm-eating Warbler. It is said to be much more varied and musical than the ordinary song, though lacking in strength. It is given as the bird moves through the woods in level flight, not while rising above the treetops as the Ovenbird and other flight singers do.

Breeding Worm-eating Warblers are drawn to wooded, leaf-covered slopes. Robert M. Mengel wrote that "except in extensively cleared areas, no steep-sided ravine or deeply shaded slope in eastern Kentucky seems to be without a pair of Worm-eating Warblers."[1] In Pike County, Kentucky, Mengel found the species to be second only to the Hooded Warbler in abundance. A typical nest is buried in a drift of leaves at the base of a bush or sapling. The location is similar to places chosen by Black-and-white Warblers.

As a ground-nesting bird, the Worm-eating Warbler faces a number of likely predators such as mice, squirrels, chipmunks, dogs, skunks, and snakes. However, the percentage of loss is probably higher from parasitization by the Brown-headed Cowbird. Friedmann regards this species as a rather uncommon host, but in some areas of abundance,

[1] Mengel, Robert M., *The Birds of Kentucky,* Ornithological Monographs no. 3, American Ornithologists' Union, 1965, p. 392.

Breeding Worm-eating Warblers are attracted to wooded, leaf-covered slopes where nests are well hidden on the ground in a drift of dead leaves.

reports of such incidents are numerous. For example, Samuel S. Dickey found that nine of twenty-two nests near his home in Greene County, Pennsylvania, contained cowbird eggs.

Although a relatively southern warbler, the Worm-eating has extended its breeding range northward during recent years. In winter, a few remain in southern Florida, but most migrate to Mexico, Central America, Panama, and the West Indies.

In 1789, Johann Gmelin in his edition of *Systema Naturae,* the great catalog of species, included the Worm-eating Warbler from a description given by Edwards. Early American ornithologists knew very little about the species. Wilson and Audubon never saw its nest.

Audubon's description, undoubtedly from hearsay, is entirely wrong: "a structure of moss and catkins, placed in a woodland shrub." His color plate depicts a male and female in a spray of common pokeberry, which, Audubon incorrectly declared, was the birds' favored food.

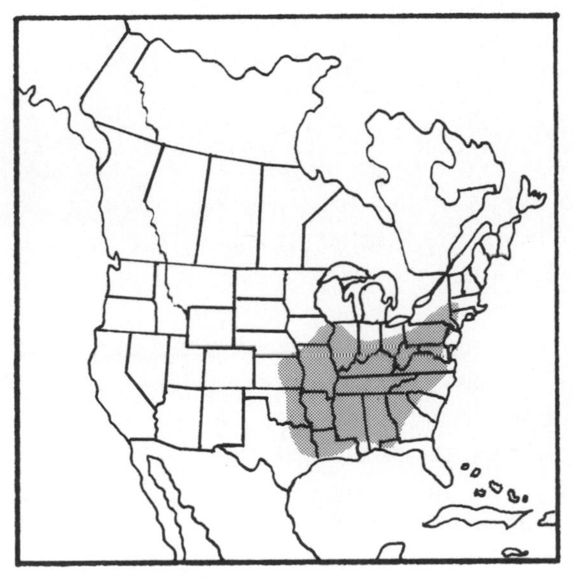

Breeding Range of Worm-eating Warbler

37. Swainson's Warbler

Limnothlypis swainsonii

PLATE 18

Many years after its discovery by John Bachman in South Carolina in 1832 or 1833, ornithologists learned a startling fact about the breeding range of Swainson's Warbler. It was long thought to be a bird of swamps and floodplain forests of the coastal plain only. Then unexpectedly it was found commonly nesting in local areas of the southern Appalachians at elevations up to three thousand feet.

Although the coastal and mountain groups are separated with no connecting populations, individuals of the quite diverse environments show no morphological differences. It has been suggested that the mountain population resulted from a relatively recent extension of range and that not enough isolation time has elapsed to modify the species.

The early history of Swainson's Warbler is also unusual. After Bachman collected five specimens along the banks of the Edisto River in South Carolina, the bird "disappeared" from science for fifty years. In 1884, William Brewster and Arthur T. Wayne resumed collecting and research, and Wayne reported the first nest near Charleston, South Carolina, on June 6, 1885.

The two isolated groups choose entirely different environments for nesting: wooded canebrake swamps in southeastern coastal plains; and in the mountains, wooded ravines and thickets of rhododendron and laurel. In the former, the birds commonly nest in cane or palmetto two to ten feet above ground; in the highlands, nests are built at the same height but in shrubs, small trees, masses of vines, briers, rhododendron, and laurel.

Males arrive first on their breeding territories after wintering in Cuba, Jamaica, Mexico (principally Yucatan), and Central America

In the mountains of the southern Appalachians, Swainson's Warblers nest in wooded ravines in thickets of laurel and rhododendron.

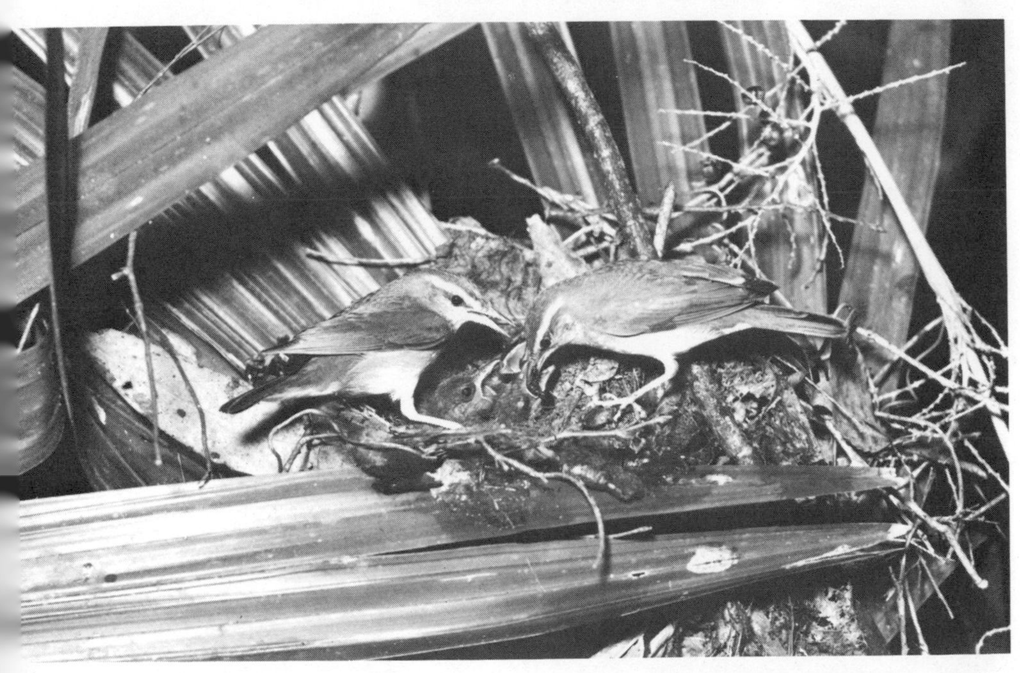

In the southern coastal plain, Swainson's Warblers are denizens of the wooded canebrake swamps, where they commonly nest in cane or palmetto.

(Honduras). Swainson's is the last of the breeding warblers to arrive in the Dismal Swamp in spring. Nesting begins soon after pair formation, usually in late April or early May. The female alone builds a bulky nest composed principally of dry leaves, mosses, and pine needles. The lining is usually of fine grasses, but Brooke Meanley tells me of six nests that he found near Macon, Georgia, and fifteen in

Except for the extremely rare Bachman's Warbler, Swainson's is the only Wood Warbler that lays white unmarked eggs. The female alone builds a bulky nest, principally of dry leaves.

Samuel A. Grimes of Jacksonville, Florida, found a Swainson's so tame that he could touch the birds at the nest.

A typical southern swamp is a common nesting site for Swainson's Warblers.

the Dismal Swamp, Virginia, that were all lined with the stems of red maple fruits or keys.

The three or four eggs are white. Except for the extremely rare Bachman's Warbler (*Vermivora bachmanii*), Swainson's is the only Wood Warbler in the United States that lays pure white eggs.

Samuel A. Grimes of Jacksonville, Florida, writes: "In each of four occupied nests that I have found, the mother bird would sit, whether incubating or brooding young, until touched. In one case, the bird would not leave until pushed off, and when I held my hand over the nest she straddled my fingers in trying to get back onto it."[1]

Swainson's Warbler has been referred to as a limited host to the Brown-headed Cowbird. Meanley tells me that he found all of four nests in Arkansas parasitized, but none of fifteen nests in the Dismal Swamp. He believes that late nesting of Swainson's in the Dismal accounts for the absence of cowbird eggs in the nests.

It has been my experience that the male Swainson's Warbler avoids the immediate vicinity of the nest except when he brings food to the young. At a nest near Charleston, West Virginia, which I watched for twelve consecutive hours, the male never sang within one hundred yards. It appeared to me that this nest was at the edge of the

[1] Grimes, Samuel A., "Injury Feigning by Birds," *The Auk,* vol. 53, no. 4, (1936), p. 478.

male's territory. The area he patrolled was evidently for feeding, for the female flew there when she left the nest. She fed with her mate but always returned to the nest alone. His visits with food for the young were silent and swift.

To me, his song is like that of the Louisiana Waterthrush—a series of loud ringing whistles. I found a ventriloquistic quality which made the male's exact location difficult to find.

For those searching for Swainson's Warbler nests in southern swamps, Alexander Sprunt, Jr., issued this warning: "Be prepared for rough going, wet and muddy clothes, and the definite possibility of encountering such undesirable swamp dwellers as the cotton-mouth moccasin, to say nothing of mosquitoes and their ilk."[2] For mountain searchers, timber rattlesnakes would replace cottonmouths.

After Bachman discovered this species, his famous colleague, John James Audubon, named it for a friend, William Swainson, a brilliant English naturalist. He named it *Silvia swainsonii.* The present generic name, *Limnothlypis* (lim-no-THLIP-iss) is from the Greek, *limne,* meaning "marsh," and *thlypis,* a small bird or finch.

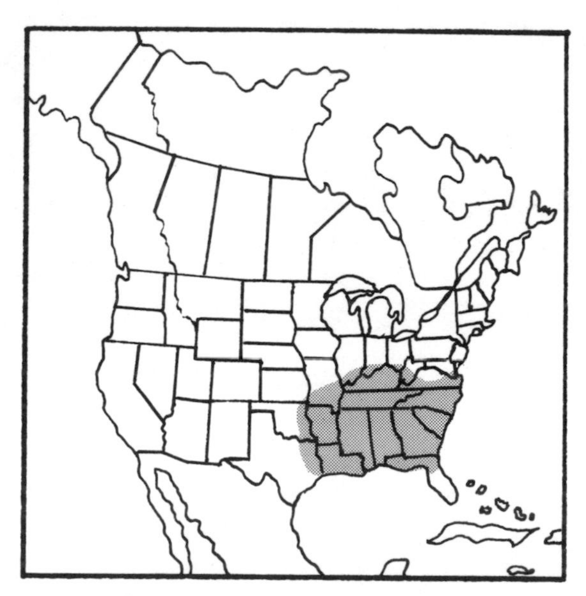

Breeding Range of Swainson's Warbler

[2]Griscom and Sprunt, p. 151.

38. *Ovenbird*

Seiurus aurocapillus

PLATE 19

Although I have seen many Ovenbird nests, rarely have I found one by actually setting out to look for it. Usually I have been walking through a leaf-strewn woods with some other purpose in mind and have accidentally flushed the tight-sitting female as she incubated or brooded. The female Ovenbird sits so tightly on her nest that she flushes only when almost stepped upon. I know of some that have been so reluctant to leave that they have been accidentally trampled. The behavior of the surprised female has always been the same—she flutters from the nest with wings and tail dragging as though she were mortally wounded. I daresay this "broken wing act" often works in luring a predator away.

The Ovenbird was described by Linnaeus in 1766 and named *Motacilla aurocapilla* and considered one of the Old World wagtails. Dr. John Latham, an English zoologist, transferred it to *Turdus aurocapillus* and called it a Golden-crowned Thrush. It does suggest a small thrush with its olive green back and black-streaked underparts, but the breast is streaked, not spotted like a thrush's breast. Swainson gave it the present generic name *Seiurus* (sigh-YOU-rus) from the Greek *seio* meaning "to shake or move to and fro" and from *oura,* "tail." The species name *aurocapillus* (aw-roh-kah-PILL-us) is Latin from *aurum,* "gold," and *capillus,* "hair," referring to the bird's orange crown.

The Ovenbird gets its English name from the shape of its nest, which is carefully disguised on the ground under dead leaves. Like an old-fashioned Dutch oven, the nest is shallowly domed over the top with the opening in the side. The entrance is often a mere slit in the grasses and dead leaves used in construction.

Ovenbird nests are usually found by accidentally flushing the tight-sitting female. When surprised, the female flutters from the nest with wings and tail dragging.

During eighteen attempts at dawn during one season in Pennsylvania, Russell T. Norris and I were successful on three occasions in photographing Brown-headed Cowbirds laying eggs in nests of a Song Sparrow, Red-eyed Vireo, and Ovenbird. At the Ovenbird's nest, the cowbird had removed one of the warbler's two eggs the day before. Knowing that this was typical of the cowbird (to remove an egg, making room for her own), Norris and I were in a blind a few feet from the Ovenbird's nest before first light on June 1. At exactly 4:58 A.M. a female cowbird flew to the ground in front of the Ovenbird's nest and entered immediately. As soon as she settled down to lay her egg, I flashed the picture that accompanies this chapter.

The Ovenbird is a very common victim of the cowbird. As many as seven and eight eggs of this parasite have been found in a single nest. Dr. Harry W. Hann, who made a life history study of the Ovenbird in Michigan, found from one to four cowbird eggs in twenty-two of forty-two nests with eggs. In one summer, I followed seven nests of the Ovenbird, six of which were parasitized. A total of twelve cowbird eggs were laid in those six nests.

For her nest, the Ovenbird chooses the leaf-covered floor of a deciduous woods, especially one with low undergrowth.

The author caught the Brown-headed Cowbird in the act of laying her egg in an Ovenbird's nest at dawn in a Pennsylvania woodland.

The song of the male Ovenbird is interpreted by everyone as *teacher, teacher, Teacher, TEACHER,* in a powerful voice, loud and ringing. The song consists of two-syllabled phrases that increase in volume to the end. Males vary in their emphasis on the *teacher* phrases; some render it as *TEACH-er;* others give it as *teaCHER.* The Ovenbird also has a flight song that is more or less a wild jumble of musical notes. There is at least one accepted record of a singing female.

During a period of forty-eight hours of observation, a male Ovenbird with young in the nest sang 215 times, a little more than once every thirteen minutes. Henry Mousley, who made this observation, compared it to a Yellow Warbler that sang 1,800 times in twenty-nine hours, or more than once every minute.

The strange sight of an Ovenbird and a Worm-eating Warbler feeding four nestling Ovenbirds in an Ovenbird's nest in Bergen County, New Jersey, was observed and recorded photographically by Stanley J. Maciula. On one occasion both birds brought food at the same time,

From the shape of its nest, concealed among dead leaves, the Ovenbird gets its English name. Like an old-fashioned Dutch oven, the nest is domed over the top.

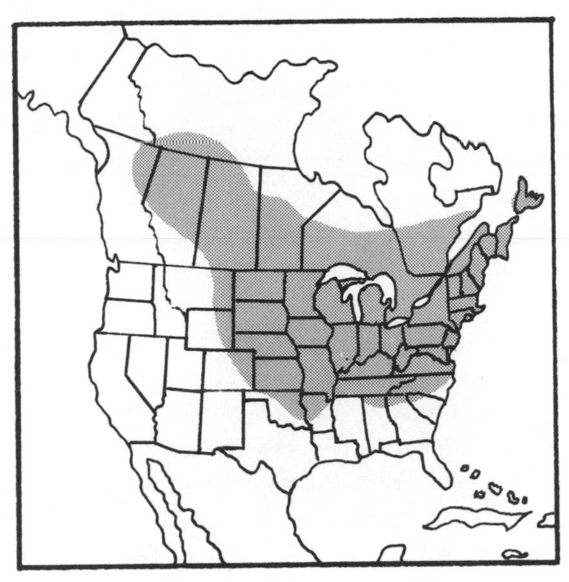

Breeding Range of Ovenbird

each feeding a different nestling. Both birds removed fecal sacs. Neither appeared hostile toward the other.

Although many Ovenbirds winter as far south as northern South America, I have found them fairly common throughout the winter in southern Florida. One January day, while eating lunch at a picnic table in Everglades National Park, I observed an Ovenbird approaching our table. Without any fear or hesitation, the bird *walked* around under the table picking up food from the ground. This bird, so shy in its summer home in the northern woods, moved unhesitatingly around our feet.

In his classic study of the Ovenbird's life history, Dr. Hann revealed some interesting facts. The first male Ovenbirds arrive (in Michigan) from nine to fourteen days before the first females. The average population is about one pair of birds to each 1.2 hectares (3 acres). The nest is built by the female, and the male does not come to it often until the eggs hatch. Nests are found to contain hair as a lining. The hair is the last material to be added and its presence indicates a finished nest. The first Ovenbird's eggs are laid on the first, second, or third morning after the lining of hair is placed in the nest.

First nests nearly always have five eggs, and subsequent nests from three to five. Dr. Hann observed two cases of bigamy, and in both cases the male took on an extra mate while the first female was incubating.

After hatching takes place the female eats the shells. Eggs that fail to hatch are left in the nest. Some young leave the nest when seven days old. At eight days old, the young are ready to leave if they have not already done so.

Dr. Hann attributed the loss of many eggs and young to the predations of red squirrels. Strangely enough, with only one exception, he found the growth of the Ovenbirds in parasitized nests to be approximately equal to that in cowbird-free nests.

In only one instance did Dr. Hann find a bird raising a second brood, after raising a part of the first brood successfully. In nearly every case, adult Ovenbirds disappeared from the woods as soon as the young were old enough to care for themselves.[1]

[1] Hann, Harry W., "Life History of the Oven-bird in Southern Michigan," *Wilson Bulletin,* vol. 49, (1937) pp. 146–237.

39. *Northern Waterthrush*

Seiurus noveboracensis

PLATE 22

Early one June morning, my friend Ralph (Bud) Long worked his way into an alder swale near his camp on Mount Desert Island, Maine, determined to find the nest of a Northern Waterthrush that he was positive was hidden there. Five hours later, he emerged after following a pair of these elusive birds from sphagnum hummocks, to upturned tree roots, to rotted stumps, to cinnamon fern clumps, all to no avail. If he had ever been near a nest, he didn't know it.

At his woods cabin at the edge of the swale, he suddenly discovered that he had lost his pipe. Then he remembered that he had laid it on a stump next to a root upon which he had sat while fighting off mosquitoes and black flies. He fought his way back through the water and tangled alders to the stump, where he found his pipe. As he turned to leave, a Northern Waterthrush darted from a cavity in the upturned tree root upon which Bud had rested. Deep inside the cavity, completely hidden behind mud and dangling roots, was a nest and five eggs. It is shown in one of the photographs that illustrate this chapter.

The location of Bud Long's nest, secluded in an upturned root of a fallen tree, is typical. In two weeks, James Bond and James Gillen found six nests in a rhododendron swamp in the Pocono Mountains of Pennsylvania, all in upturned tree stumps. In his life history study of this species, Stephen W. Eaton writes: "Fifteen nests found during this study ranged from 7.5 cm. to 0.6 meters above water. They were most often found within 0.3 meters of the water, tucked in behind rootlets of trees which had fallen over leaving their root systems at right an-

A Northern Waterthrush commonly chooses the root of an upturned tree, usually with water below it, as a hiding place for its nest. Bud Long found this one while looking for his lost pipe.

gles to the water surface. Directly beneath the nest, in such situations, there is usually a large pool of water, formed by the removal of the mass of roots and surrounding muck."[1]

The Northern and Louisiana Waterthrushes resemble each other in appearance and the way they move, but the breeding habitats of the two are decidedly different. There is no competition between them for food or nesting sites. John Bull put it succinctly when he wrote, "The Northern Waterthrush is to the wooded *swamp* what the sibling Louisiana Waterthrush is to the wooded *stream*."[2]

The Northern Waterthrush winters in Mexico and the West Indies south to Panama and northern South America. It is abundant in Pan-

[1] Eaton, Stephen W., *A Life History Study of Seirus noveboracensis,* vol. 19 of *Science Studies,* (St. Bonaventure, New York: St. Bonaventure University, 1957), p. 15.

[2] Bull, John L., *Birds of New York State* (New York: Doubleday, 1974), p. 509.

Unlike its close relative, the Louisiana Waterthrush, the Northern Waterthrush is a swamp dweller, not a streamside nester.

When leaving the nest, the female Northern Waterthrush slips off slowly, quietly and deliberately. She *walks* with shortened steps, head down and breast close to the ground.

ama, where it defends feeding areas in mangrove swamps on both the Atlantic and Pacific coasts. This species remains on its winter grounds for at least six months.

Ordinarily the Northern Waterthrush arrives in its breeding range two to three weeks later than the Louisiana Waterthrush. Flight songs are frequent for the Northern, rare for the Louisiana. Singing is not suppressed after pair formation in the Northern; it often ceases or becomes infrequent with the Louisiana. The female alone builds the Northern Waterthrush's nest; both sexes help to construct the Louisiana's nest.

The Northern's song, like the Louisiana's but not so thrilling, is a loud high-pitched whistle that carries far. It has been described as a hurried musical two- or three-parted rendition that is well written as *hurry, hurry, hurry, pretty, pretty, pretty.* Like the Ovenbird, the Northern Waterthrush has a flight song that sometimes carries the bird high above the treetops.

During the twelve days that the female incubates, the male patrols his territory and sings from trees usually at some distance from the nest. Eaton describes behavior during this period:

"When leaving the nest, the female slips off slowly and deliberately. She walks with shortened steps, head down and breast close to the ground, reminiscent of a mouse. When about 10 meters away she gets up from her 'haunches,' stretches and starts feeding. Her approach to the nest is similar. If she has been quite far in her search for food, she will fly to a point about 10 meters from the nest and walk mouse-like the rest of the way."[3]

In Maine and Pennsylvania, where I have studied and photographed Northern Waterthrushes, they have disappeared from their nesting grounds by the end of June. If my observation is typical, this species must spend less than three months in its northern habitat. To check my observation, I wrote to Robert Leberman, resident bird bander at the Powdermill Nature Reserve field station of the Carnegie Museum of Natural History, at Rector, in the Pennsylvania mountains. Since the Northern Waterthrush does not nest in that area, any fall records gathered by Leberman would be migrant birds.

In 1983, Leberman banded his first migrant Northern on July 8, the earliest fall date he had ever recorded. Coincidentally, I saw my last

[3] Eaton, p. 17.

Northern Waterthrush in Pymatuning Swamp in northwestern Pennsylvania on June 25, 1983. Other fall migration arrival dates from previous years furnished me by Leberman are: August 4, 1974; August 7, 1975; August 11, 1976; July 29, 1977; July 25, 1978; August 1, 1979; August 5, 1980; July 26, 1981; and July 23, 1982.

The species name, *noveboracensis* (no-veh-bore-ah-SEN-sis), is Latin for "of New York." New York is the type locality for the species. Other former names include New York Warbler, Northern Small-billed Waterthrush, and Water Wagtail. The waterthrush was included in Gmelin's 1789 edition of *Systema Naturae.* Gmelin added this species on the basis of a description by Dr. John Latham.

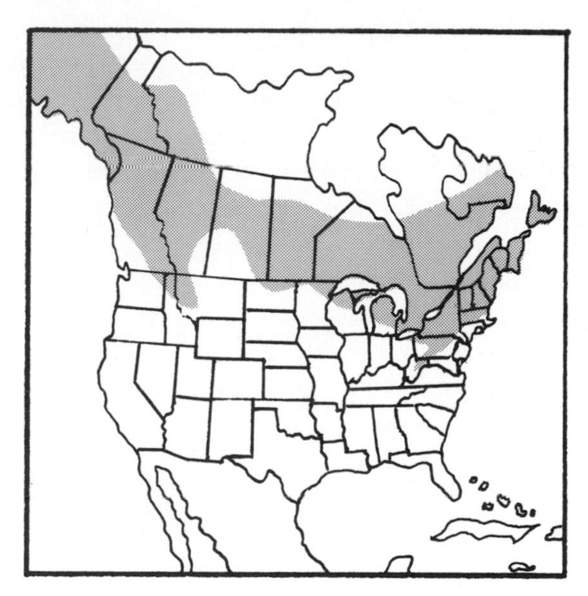

Breeding Range of Northern Waterthrush

40. *Louisiana Waterthrush*

Seiurus motacilla

PLATE 22

I can get very nostalgic at the mere mention of the name waterthrush. It conjures up the memory of spring in Pennsylvania—hillsides of trillium, skunk cabbage in the bottomlands, peepers calling from the grass, scarlet cups, mourning cloaks, water rippling over rocks, and the wild ringing call of a Louisiana Waterthrush as he races up and down Little Buffalo Creek. This song, more than that of any other warbler, exemplifies wilderness. It is one song that I can truthfully describe as exciting.

This is one of the first spring warblers to return to us. Look for it at the same time the Black-and-white Warbler appears. Its winter home is in Mexico and Central America, and, although it has much farther to travel than some other warblers (e.g., Yellow-rumped, Pine), it arrives in western Pennsylvania in mid-April or before. Likewise, the fall departure is characteristically early, most having left by late August.

I was fortunate to discover the nest of a Louisiana Waterthrush the day the pair started to build. It was on the side of a narrow ravine about six feet up a steep bank above the edge of a small brook. A hollow had already been dug in the moist earth under a slightly overhanging ledge of small rocks, moss, and roots. Both adults were busy carrying wet leaves to fill the hollow. When I left late in the day, the entrance to the cavity was almost blocked with leaves. Three days later the nest was completely lined and ready for eggs. Six days after that the female was incubating six eggs. All hatched exactly thirteen days later. The male assisted in feeding the young, but I never saw him incubate or brood. During the time the female incubated, the male sang infrequently, a sharp contrast to his persistent singing ear-

The nest of the Louisiana Waterthrush is placed in a steep bank above a stream of running water. Its close relative, the Northern Waterthrush, is a swamp nester.

lier. When I arrived at the nest on the tenth day after the young had hatched, one nestling had fledged. When my hand touched the backs of the five piled high in the nest, they exploded in all directions. The parents became frantic in their excitement while luring the young away.

A unique feature that has been an aid to me in finding Louisiana Waterthrush nests is the long walkway or platform of leaves that the birds build in front of the nest, often leading toward the stream below. Stephen W. Eaton offers another suggestion for finding a nest when the young are being fed. "When the young are about four days

The territory of the Louisiana Waterthrush is long and narrow and follows the course of a swift-moving stream in a wooded ravine.

old the nests were located easily if water in the stream was low. The parents left fecal sacs in the stream within a few feet of the nest."[1]

Three of the eight nests I have found were parasitized by the Brown-headed Cowbird. Two held two cowbird eggs; one contained one. Of the other five nests, three had five eggs and two had six. In Eaton's exhaustive study of this species in the Ithaca, New York region, nine of sixteen nests (56 percent) were parasitized by the cowbird. These nests contained from one to four cowbird eggs each, with a total of eighteen. Twelve young cowbirds were raised to leave the nine nests.

In his study, Eaton found that early in the season the Louisiana Waterthrushes feed entirely in the stream. He writes, "Here they can be seen flipping leaves and picking up the disturbed animals beneath, poking their bills into crannies or picking things directly out of the water from a wading approach."[2] To this I can add that at nests where

[1] Eaton, Stephen W., "A Life History Study of the Louisiana Waterthrush," *Wilson Bulletin,* vol. 70, no. 3 (1958), pp. 211–212.

[2] Ibid., p. 219.

I have watched adults gathering food for their young, the old birds often leaped into the air from the edge of the stream to snare flying insects. At a nest I studied in 1983, most of the insect food brought to the nestlings was adult or larval mayflies.

The territory of a Louisiana Waterthrush is long and narrow and follows the course of a swift-moving stream in a wooded ravine. In favorable habitat, Eaton found nests as close as two hundred meters apart.

The Louisiana Waterthrush, formerly called a Large-billed Waterthrush or Water Wagtail, was often confused by early ornithologists with its close relative, the Northern Waterthrush, then known as the Small-billed Waterthrush. Neither Wilson nor Nuttall recognized that there were two species.

The species name, *motacilla,* means "tail mover." Both scientific names allude to the bobbing up and down of the bird's head and the foreparts of its body with a springing motion of its legs, like a Spotted

This Louisiana Waterthrush arrived at its nest with a beak filled with larval and adult mayflies as food for its young.

Sandpiper. The Waterthrush also raises and lowers its tail as it walks. The name *Louisiana* refers to the old French territory, not the present state. The name thrush, of course, is a misnomer. Louis Jean Pierre Vieillot, a French contemporary of Dr. John Latham, is given credit for being first to describe the Louisiana Waterthrush in 1808.

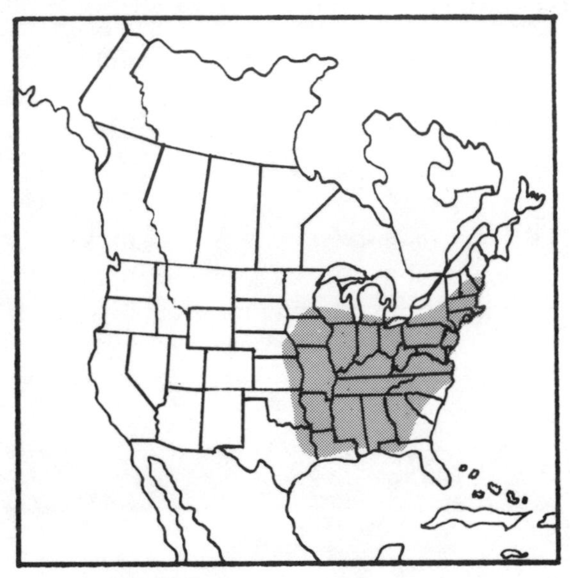

Breeding Range of Louisiana Waterthrush

41. *Kentucky Warbler*

Oporornis formosus

PLATE 9

The greatest enemy of the Kentucky Warbler in the mountains of Pennsylvania is the white-tailed deer. Yes, deer! This startling revelation was made to me by Robert C. Leberman.

"Since I have been working with Kentucky Warblers, one thing has become clear to me," Leberman wrote to me. "The biggest limiting factor on population size in the western Pennsylvania mountains is not the cowbird, snakes, or small mammal predation; it is the White-tailed Deer.

"Over much of this mountain area the forest understory has been grazed so heavily that there is no habitat left for Kentucky Warblers . . . and presumably many other species needing heavy shrub cover," he concluded.

The condition cited here is understandable when it is realized that the Kentucky Warbler is a ground-inhabiting species. Like the Golden-winged Warbler, it places its bulky nest on or just above ground level, commonly in grass tussocks, bedstraw, or goldenrod, or in the ground forks of small saplings. One nest that I found was buried in a clump of barberry. Leberman told me that in the Powdermill area a few pairs of Kentuckys that nest in spicebush and greenbrier usually survive because deer do not normally eat these plants.

Leberman's comments are concerned with Kentucky Warblers near the northern extremity of their breeding range. Most pairs nest south and west of the Allegheny Mountains.

Like some other ground-inhabiting species (Ovenbird and waterthrushes), the Kentucky does not hop, it *walks,* often tilting its body and wagging its tail. One of the early English names for it was Ken-

In western Pennsylvania mountains, deer have limited the population of Kentucky Warblers by heavily grazing the understory and thereby eliminating nesting sites.

The author prepares to photograph the nest and eggs of a Kentucky Warbler in a Pennsylvania woodland.

tucky Wagtail. The birds forage on or near the ground, where they search for grubs, plant lice, spiders, caterpillars, and other insect larvae. Overturning leaves in search of food is characteristic of the Kentucky; leaping from the ground to snare a low-flying moth is another habit.

The male Kentucky Warbler forsakes his low surroundings to pour out a clearly whistled vibrant song of two syllables repeated four or five times. Those who know this loud song intimately are almost unanimous in the opinion that it resembles the chortling of a Carolina Wren. The *tur-dle tur-dle tur-dle tur-dle tur-dle* carries far, ringing through the spring woods. The bird is a persistent singer. Dr. Chapman watched a male sing 875 times in three hours. Allowing for a few brief intervals, the songs came every twelve seconds.

Look for this dainty woodland bird in damp stream bottoms and ravines, deep-shaded and dense thickets, woodland borders, and

Look for the Kentucky Warbler's nest in damp bottomlands and ravines, deep-shaded and dense thickets, woodland borders, and swamp openings.

Like the Golden-winged Warbler, the Kentucky places its bulky nest on or just above ground, commonly at the base of a sapling or in a grass tussock, bedstraw, or goldenrod.

Alexander Wilson, who discovered the Kentucky Warbler, named it for the state where he found it. It winters from southern Mexico to Colombia.

swampy openings. I found one female incubating four eggs in a leafy nest she had constructed in a soggy skunk cabbage opening in a deciduous woods.

The Kentucky Warbler winters from southern Mexico to Colombia. It is an abundant winter resident in the lowlands of Panama.

Alexander Wilson, who discovered this species, named it for the state in which he found it abundant. The generic name *Oporornis* is Greek for "autumn bird"; *formosus* is Latin for "beautiful."

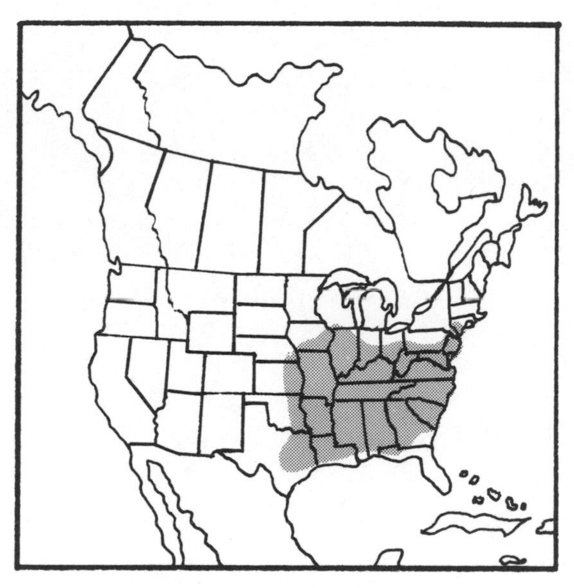

Breeding Range of Kentucky Warbler

42. *Connecticut Warbler*

Oporornis agilis

PLATE 24

If you live in the Atlantic coastal states and wonder why you never see Connecticut Warblers during spring migration, or if you live in the Mississippi Valley and wonder why you never see Connecticut warblers during fall migration, it is because of the routes the birds travel. They leave their winter home in Venezuela and Brazil in the spring and head north to the Florida peninsula where they change course and fly diagonally westward to the Mississippi basin. A flight up the Mississippi Valley takes them to their breeding grounds. In the fall, Connecticut Warblers fly eastward a thousand miles to New England and thence south along the Atlantic Coast to Florida, the West Indies, and South America. Because individual birds don't always follow tra-

One of the Connecticut Warbler's favorite nesting habitats is the great tamarack, black spruce, and muskeg bogs common in north-central United States and southern Canada.

Cerulean Warblers, male, left; female on nest

Swainson's Warblers

Wilson's Warbler

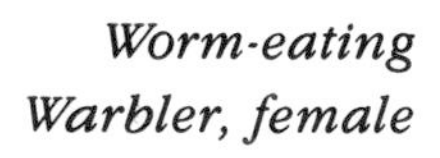

Worm-eating Warbler, female

PLATE 18

Hooded Warblers, male, right; female, left

Ovenbird approaching ovenlike nest

Male Common Yellowthroat

Male Prothonotary Warbler

Male Black-and-white Warbler

Red-faced Warbler

Painted Redstart

Louisiana Waterthrush

Northern Waterthrush

Female Olive Warbler

Yellow-breasted Chat

Canada Warbler

Male Mourning Warbler

Male Connecticut Warbler

Male MacGillivray's Warbler

These fledgling Connecticut Warblers will migrate along the Atlantic Coast to their winter home in the West Indies and South America. In the spring, they will return via the Mississippi valley.

ditional patterns of migration, bird watchers occasionally will see a Connecticut Warbler in spring in the East, or in fall in the West. However, such instances are considered rare and surprising.

The Connecticut deserves its reputation for being one of the most elusive of Wood Warblers. Once it reaches its breeding grounds it disappears into dry, open poplar woods, or the great tamarack, black spruce, and muskeg bogs that are common in the north-central United States and adjacent southern Canada. The Connecticut Warbler hides its nest on the ground in a sphagnum hummock and occasionally on higher dry ground. It is always well concealed by surrounding vegetation.

Compounding an observer's problems in hunting for a nest is the female Connecticut's habit of landing thirty to forty feet away from it and walking quietly through the underbrush to the spot. The male is no help either; his singing has little to do with nest location.

The female Connecticut Warbler's habit of landing thirty to forty feet away from the nest and walking quietly through the underbrush makes finding the nest difficult.

Richard C. Harlow, who found eight nests near Belvedere in Alberta, Canada, described the habitat there as dry well-drained ridges and poplar woods. Harlow pointed out that the Connecticut builds a much frailer nest than its close relative, the Mourning Warbler. Typically, it is a deep rounded cup on a foundation of leaves or sunk in moss. It is lined with grasses.

If it were not for the clear, ringing voice of the male, few bird watchers would ever see a Connecticut Warbler in the swampy wilderness or in the many-branched poplar trees where it spends its summers. Even with the song as a guide, the observer will usually have difficulty in locating the singer, who is adept at hiding. In addition, the song has a ventriloquial quality. Ernest Thompson Seton, who first found the nest of this species in Manitoba in 1883, interpreted the song as *beecher-beecher-beecher-beecher-beecher-beecher.* Seton described it as "like the song of the Golden-crowned Thrush [Ovenbird] but differs in being in the same pitch throughout instead of beginning

The Connecticut Warbler's nest is a deep rounded cup on a foundation of leaves or sunk in moss. It is always well concealed.

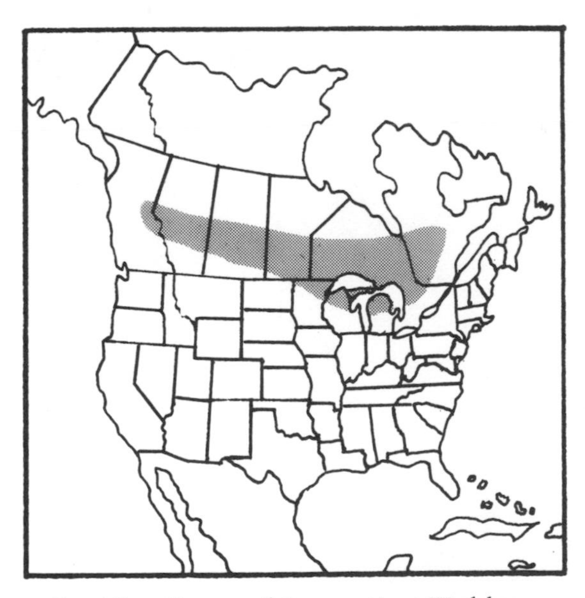

Breeding Range of Connecticut Warbler

with a whisper and increasing the emphasis and strength with each pair of notes to the last.''[1]

Betty and Powell Cottrille, who have studied the Connecticut on its breeding grounds in Michigan and in Minnesota, tell me that the male's song in those areas is more in triplets than Seton and others say. They like the interpretation *tu-chibee-too, chibee-too, chibee-too.* The call or alarm note is *whik.*

Alexander Wilson discovered this species and notes about it were published in 1812. He gave it the English name of Connecticut Warbler in honor of the state where it was first found. The species name *agilis* (AH-jih-liss) is Latin for ''active'' or ''nimble.''

[1] Bent, ed., p. 519.

43. *Mourning Warbler*

Oporornis philadelphia

PLATE 24

Bird photographers will have empathy for me in my experience at the nests of two pairs of Mourning Warblers in the Cheat Mountains of West Virginia. At one nest, on the banks of Shaver's Fork of the Cheat River, I spent almost two days watching a female feed young while her mate refused to come within twenty feet of the nest and the camera. Later that week, I found a second nest on a hillside along the road up to Gaudineer Knob. At this nest, the male fed regularly and fearlessly; the female appeared on several occasions and then disappeared for the rest of the day.

The Mourning Warbler winters from Central America south to Ecuador. In spring migration it moves north through Texas and the Gulf states and then spreads out to occupy an extensive breeding range. The species migrates rather late in the spring compared to many other wood warblers (arrives in West Virginia and Pennsylvania during the second and third weeks of May). The return journey south begins in July. Immature birds have been seen in Texas as early as August 4.

Mourning Warblers occupy a number of dissimilar habitats, including slashings, dry, brushy clearings, roadside tangles, swamp thickets, dense second growth, and aspen-birch woods. Nests are always on or near the ground. A nest thirty inches above ground, reported by T. S. Roberts, may be the highest known. Typical nests are well concealed in tangles of briers or herbaceous plants (jewelweed, ferns, goldenrod, grass tussocks). The exterior is bulky, made of dry leaves, vine stalks, coarse grasses, and bark pieces. The lining is usually of grasses, fine rootlets, and animal hair.

The Mourning Warbler winters from Central America south to Ecuador. Compared to many other Wood Warblers, it migrates late in the spring. After nesting, it moves south as early as July.

In addition to roadside tangles, Mourning Warblers occupy a number of dissimilar habitats: slashings, dry brushy clearings, swamp thickets, dense second growth, and aspen-birch woods.

While watching nest behavior, I discovered that the female Mourning Warbler rarely flies directly to the nest but lands several yards away and silently creeps through the vegetation. When the young hatch, the male generally assists in feeding them.

None of the four nests I have found was parasitized. Each contained four warbler eggs. Friedmann agrees that the Mourning Warbler is generally an uncommonly utilized host of the Brown-headed Cowbird.

The song of the male has been variously described as resembling the songs of the Common Yellowthroat, Kentucky Warbler, Ovenbird, Northern Waterthrush, House Wren, and Lincoln's Sparrow. Let me add that to my ears it sounds like the song of a Carolina Wren or a Kentucky Warbler. It is loud, musical, and usually in two parts. Most bird watchers agree that Roger Tory Peterson's description of the male's song is a good one: *chirry, chirry, chorry, chorry* (*chorry* lower in pitch).

Whether the Mourning Warbler and MacGillivray's Warbler (*O. tolmiei*) are two different species or races of the same species has been

Mourning Warbler nests are always on or near the ground. They are concealed in tangles of briers or herbaceous plants.

Mourning Warbler eggs hatch after twelve days of incubation. The nestlings leave the nest in eight or nine days. These fledglings are still dependent upon their parents for food.

the subject of considerable debate among ornithological taxonomists. That the two hybridize is undisputed. Their breeding ranges overlap in central Alberta, Canada, but most of the probable hybrids have been migrants collected outside that area.

The first recorded indication of hybridization was reported by George W. Cox in 1973, from a specimen taken in Alberta in 1963. Additional possible hybrids, all migrants, reported since then have added to the uncertainty. Some of the supposed hybrids have proven, upon further study, to be slightly aberrant individuals of one or the other species.

Wesley E. Lanyon and John Bull declare that "these two forms are, in fact, extremely closely allied and may well be conspecific."[1] George A. Hall has made an extensive study of the situation to date. He explained his views to me in a letter: "You can see by my com-

[1] Lanyon, Wesley E., and John Bull, "Identification of Connecticut, Mourning and MacGillivray's Warblers," *Bird-banding,* vol. 38, no. 3 (1967), p. 193.

ments that I have not made up my mind on the proposition on conspecificity of these two species, although I am inclined to think they are one species."

A hybrid presumed to be between a Mourning Warbler and a Blue-winged Warbler has also been reported.

In 1832, Alexander Wilson discovered the Mourning Warbler at a marsh near Philadelphia, hence the scientific name. He called it Mourning Warbler because the black on its breast suggests a symbol of mourning, a crepe. The bird Wilson collected was the only Mourning Warbler he ever saw.

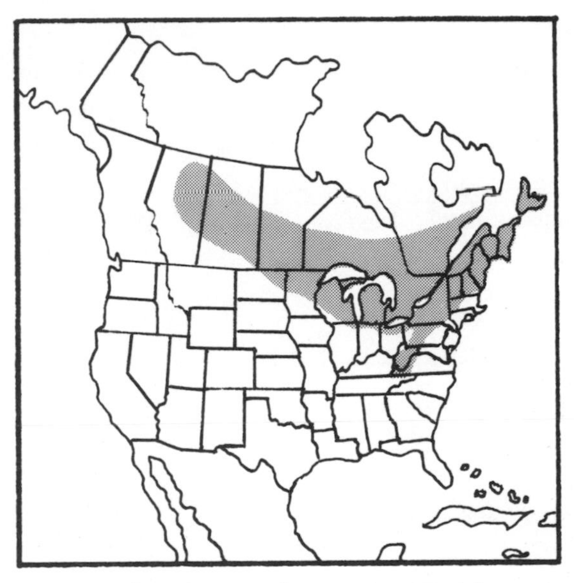

Breeding Range of Mourning Warbler

44. *MacGillivray's Warbler*

Oporornis tolmiei

PLATE 24

A highway road crew in Utah nearly ruined the only chance I have ever had of taking a picture of MacGillivray's Warbler. The nest that I was planning to photograph contained young birds and was just a few feet off the highway on the brushy shoulder of the road. I had just set up my blind and camera when two workmen in a pickup truck parked near me. They took two scythes from the back of the truck and started to clear brush near my setup. If they continued, my operation was over. So, when one of the men asked me what I was doing, I told him that the state had hired me to count rattlesnakes along the edge of the road in this location. I said that I had seen only five, but that I had not been there very long. Out of earshot, the two men held a conference. A few minutes later, they drove away. That was the last I saw of them.

From its winter home in central Mexico and southern Baja California and south to western Panama, MacGillivray's Warbler moves north in spring to its breeding area in the coniferous and mixed forests of the West, where some individuals nest as high as six thousand or more feet above sea level.

The nesting habitat of MacGillivray's is not unlike that of the Mourning Warbler: low dense undergrowth; shady, damp thickets; forest edges, burns, and brushy hillsides. The female builds her nest in one of a variety of small dense shrubs, one to five feet above ground, but most often two to three feet. The nest is loosely built of coarse grass blades and stems and bark shreds lined with soft grasses, fine rootlets, and animal hair. It is commonly held upright by several stalks of the plant onto which it is woven.

The author found a MacGillivray's Warbler's nest a few feet off this Utah highway. A road crew almost ruined chances for a photograph.

Female MacGillivray's Warblers place their loosely built nests in small dense shrubs, one to five feet above ground. Four eggs is the usual clutch.

The broken eye-ring of this nesting MacGillivray's Warbler male is a sure clue to identification. In fall migration, however, female and immature Mourning Warblers may have partial eye-rings.

The MacGillivray's song is similar to the Mourning's. In a study of the genus *Oporornis* P. B. Hofslund wrote: "The Mourning Warbler in northeastern Minnesota sounds to me more like the MacGillivray's Warbler than it does like the Mourning Warblers of the east."[1] MacGillivray's song is two-parted. The first phrase of three or four notes is delivered in a higher pitch than the last phrase. It may be interpreted as *sweet-sweet-sweet-sweet, sugar-sugar.*

Of the four members of the genus *Oporornis* in the Wood Warbler subfamily (Parulinae), three are look-alikes: Connecticut, Mourning, and MacGillivray's Warblers; the Kentucky Warbler is the exception. Most bird watchers consider the eye-ring, or lack of it, an identifying characteristic for the trio. This feature is helpful but not always reliable. The MacGillivray's Warbler has an incomplete or broken eye-ring. The Connecticut always has a complete eye-ring. But in the

[1] Hofslund, P. B., "The Genus Oporornis," *Flicker,* vol. 34, no. 2 (1962), p. 44.

Mourning Warbler, where absence of an eye-ring is one of the diagnostic points, the characteristic is variable. The most perplexing problem arises during fall migration, when Mourning Warbler females and immatures often have partial eye-rings. To add to the confusion, sometimes a female Mourning Warbler will have an eye-ring in spring.

MacGillivray's Warbler is the western counterpart of the Mourning Warbler, with which it is known to hybridize in a range overlap in Alberta, Canada. There is growing belief that the two may be conspecific. (See chapter 43.)

In 1839, John K. Townsend discovered this species and named it for his friend, William Fraser Tolmie, a doctor and officer in the Hudson's Bay Company; this included both the species name, *tolmiei* (toll-MEE-eye), and the English name, Tolmie's Warbler. Later, Audubon dedicated the same species to his assistant in Scotland, William MacGillivray, and gave it the species name *macgillivrayi.* Townsend's notice of the species preceded Audubon's, so the scientific name *tolmiei* was official. However, in time the English name lost favor and Audubon's choice, MacGillivray's Warbler, replaced the name Tolmie's Warbler.

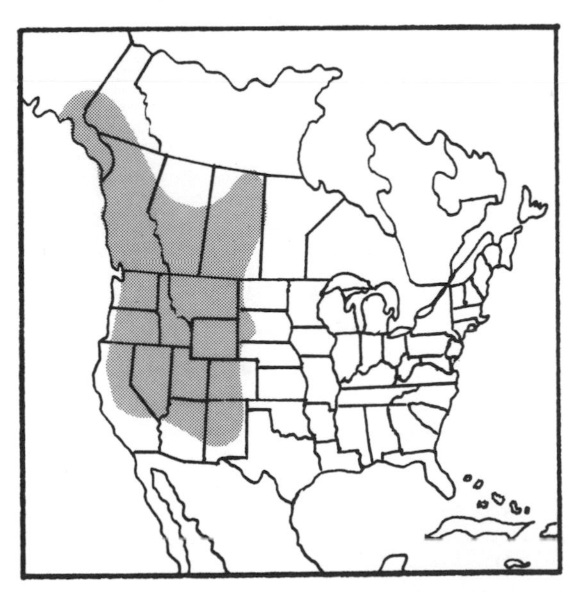

Breeding Range of MacGillivray's Warbler

45. Common Yellowthroat

Geothlypis trichas

PLATE 20

I have listened to songs of the Common Yellowthroat from Maine to Florida and across the southern and northern United States to California, and I have concluded that anyone who remembers the phrases *wich-i-ty wich-i-ty wich-i-ty wich-i-ty wich-i-ty* will have no problem identifying this abundant warbler regardless of which of the many geographic races is encountered.

The song has innumerable variations. A male I heard in Maine sounded something like a Nashville Warbler. In Pennsylvania, I made note of a male singing *whit-i-shee.* In Texas, the birds seem to say *wit-tit-ee.* In California it sounded like *wreech-ity.* In Michigan, *wheet-to.* But regardless of where I have heard these variations, I have easily recognized the voice. It has a characteristic rhythm; it is loud and clear and is much more definite and distinct than the songs of most warblers.

Donald J. Borror studied recorded songs of 411 Common Yellowthroats representing ten North American subspecies. He concluded that "a Yellowthroat song consists of from two to five repetitions of a sequence of notes known as a phrase. The songs of a given individual varied little except in length. The phrases of different birds showed considerable variation with about every third bird singing a different phrase. There was a greater variation among local populations in the east than in the west."[1]

The generic name *Geothlypis* (jee-OTH-lip-iss) means "ground

[1] Borror, Donald J., "Songs of the Yellowthroat," *Living Bird,* 6th Annual (1967), p. 160.

If this young Common Yellowthroat is a male, he will grow up to sing the characteristic *wich-i-ty wich-i-ty wich-i-ty* song of his father. The song is more distinctive than the songs of many warblers.

bird"; and the species name is *trichas* (TRY-kas), a kind of thrush. The former is appropriate, for the Yellowthroat is definitely a bird of the ground and low bushes, a skulker that moves quietly, out of sight much of the time. Nevertheless, the Maryland Yellowthroat, as it was initially known, was one of the first birds described by Latham and accepted and named by Linnaeus.

Most bird watchers have had the exasperating experience of searching a roadside tangle or a brushy meadow trying to get a glimpse of the bird with the *wich-i-ty* song. The male has an uncanny ability to remain hidden even when on the move. The female presents a still more difficult problem. Except for her scolding wrenlike alarm note her presence in shrubby areas would go unnoticed.

The Common Yellowthroat has one of the largest continuous breeding ranges of any North American warbler. To travel to and from the breeding range, Yellowthroats have what some observers call "leapfrog" migration. Individuals that live in the southern part of the species' range (Florida, Gulf Coast, Texas, and the Southwest) are almost nonmigratory. Those that breed in the North migrate to the

Both Common Yellowthroat parents feed the young, which remain in the nest for eight to nine days after hatching. These youngsters, being fed by the female, are ready to fledge.

West Indies, Mexico, and South America and literally "jump" over the permanent homes of the southern individuals.

The Common Yellowthroat is not a dooryard bird. The typical habitat is low dense vegetation, which is most prevalent in wet areas such as edges of small streams and ponds, open bogs, sedge meadows, and similar situations. It appears that the vegetation in such areas, rather than the presence of moisture, attracts the nesting birds, for I have also found Yellowthroats nesting in dense undergrowth on dry hillsides, blueberry fields, hedgerows, and brushy thickets.

Male Common Yellowthroats arrive on their northern breeding areas in late April and are joined a week or two later by females. The new arrival chooses her mate quickly and during courtship the male follows her closely. During this period his song is heard less frequently

The Common Yellowthroat has one of the largest continuous breeding ranges among Wood Warblers. Those that breed in the deep South are practically nonmigratory. In the North, males arrive in late April; females a week or two later.

but he resumes singing when nest building is finished. Males sing from low perches, but will mount higher into trees when faced with competition from other males. Occasionally a male may perform a flight song. Starting from a perch on a low shrub, he flies upward for twenty-five to one hundred feet before swooping downward to another perch. During his ascent, he utters a garbled song and a few call notes, but he is silent during his descent.

The female Common Yellowthroat selects the site and builds the nest alone. It is placed on or near the ground and is supported by surrounding vegetation. Compared to the nests of most warblers, Common Yellowthroat nests are quite bulky. They are built entirely of dry plant material. In approaching her nest, the female hops through weeds and grasses from as far as ten feet away. When surprised at the nest, she will slip off quietly, making nest-finding difficult. Most females attempt a second brood, or, if nests are lost through predation, as many as four may be attempted in one season.

Polygyny may be more common in this species than the reports

The Common Yellowthroat is not a dooryard bird. Typical nesting habitat is low dense vegetation, often in wet areas such as stream edges, bogs, and sedge meadows.

The nest of the Common Yellowthroat is on or near the ground and is supported by surrounding vegetation. The bulky structure is built entirely of dry plant material.

indicate. On Mount Desert Island, Maine, I observed one male mated to a female who later had the task of feeding four young alone when her mate abandoned her for another female in an adjoining territory. George V. N. Powell and H. Lee Jones at Huntington, New York, observed a color-banded male mated with two color-banded females, each of which successfully fledged young.

The Common Yellowthroat is a frequent victim of the Brown-headed Cowbird. Over three hundred cases of parasitism have been reported. In a Michigan study, Robert E. Stewart found ten of twenty-two nests to contain cowbird eggs; and in those ten nests, the number of cowbird eggs ranged from one to three each.

Some strange stories have been related regarding the Common Yellowthroat. For example, Austin L. Rand reported that while fishing in Florida, his party caught a three-pound bass that had the remains of a Yellowthroat in its stomach; Lucretius H. Ross told of a Yellowthroat he watched making a frantic effort to free itself from a spiderweb; an adult male was seen by Richard D. Brown in New Jersey, caught on the top of a common burdock that was in bloom; and Oscar T. Owre found a dead Chuck-will's-widow in Florida that had a male Yellowthroat wedged tightly in its throat. In the Chuck's stomach was a nearly intact male Cape May Warbler.

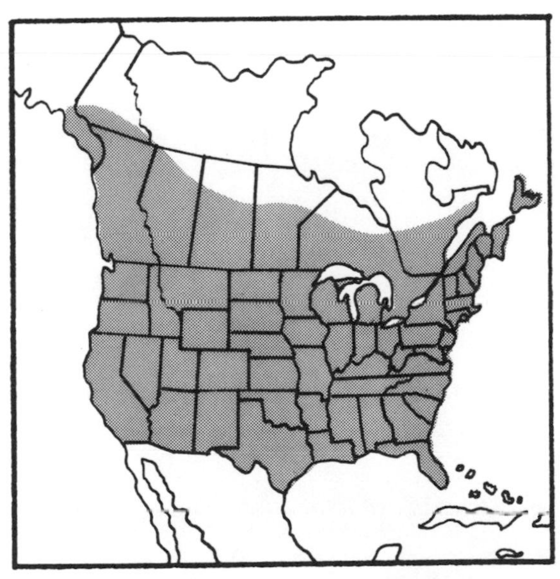

Breeding Range of Common Yellowthroat

46. Hooded Warbler

Wilsonia citrina

PLATE 19

In June 1942, during a dark period in the Second World War, my attempt to photograph a pair of Hooded Warblers at their nest caused the United States Secret Service to place my activities under surveillance for a week. The nest was in a rhododendron thicket a mile above Dam no. 9 on the Allegheny River in Armstrong County, Pennsylvania. On a Sunday afternoon, at a point thirty feet from the nest, I put up a tentlike blind where I expected to hide when I returned to photograph the birds a week later.

The following Sunday, I returned and finished my photography as planned, removed the blind, and left the area, oblivious to the consternation I had caused. Residents near the dam told me the story

My blind, erected near this Hooded Warbler's nest in Pennsylvania, was under Secret Service surveillance during World War II. (This female lacks the black head of the male.)

Hooded Warbler nests are commonly found in moist woodlands with low bushy undergrowth. Favored habitats include ravines, river bottoms, forested swamps, and dense tangles.

later. The Secret Service discovered the blind, from which they could see Dam no. 9, and they assumed that the person who erected it was spying on activities at the dam. An agent from the Secret Service had the blind under surveillance for a week, but no one came near it. Then on Sunday, when the government relaxed its vigil, the blind suddenly disappeared. I attempted to contact the Secret Service to explain, but was never able to locate the right person.

I have found nine Hooded Warbler nests, all in Pennsylvania. Five were parasitized by the Brown-headed Cowbird. As my records indicate, the Hooded is a common victim throughout its range.

This species winters in Mexico and Central America including Panama. It is rare in the West Indies. Like so many others in its family, the Hooded Warbler reaches its breeding grounds in the eastern United States in early May. Males arrive first but females are not far behind.

As soon as a territory is established, the male sings continually throughout the nesting season. He is not shy when an intruder approaches, but the female is just the opposite. She is quick to drop out of sight, and to see her, unless she is attending a nest, is difficult.

Most observers who have studied the Hooded Warbler's song are in general agreement regarding the interpretation and cadence: *ta-weet-ta-weet-eo.* The song is short, loud, and clear. To my ears, the five to eleven rapid notes are very similar to the songs of the Magnolia Warblers I hear in Maine.

In Pennsylvania I have always found the Hooded Warbler in woodland with low, bushy understory, often in moist locations such as ravines and river bottoms. Another favorite habitat is in oak-beech-maple woodland with a ground cover of herbaceous plants such as black cohosh. In the South, the Hooded is commonly found in forested swamps where dense tangles hide its movements.

Nests invariably are placed at low levels, one to six feet above ground, commonly two or three feet. They are in small bushes, herbaceous plants (cohosh is a favorite), vines, saplings, and in mountainous areas, in laurel and rhododendron. In the South, cane and palmetto are often used. Nests are neat and compact.

Nests of the Hooded Warbler are invariably placed at low levels, one to six feet above ground, typically in small bushes, herbaceous plants, vines, rhododendron, cane, and palmetto.

Field guides describing the female Hooded as lacking the black hood of the male are not altogether correct. Here is a pair of Hooded Warblers at the same nest showing the female (bottom) with a hood almost as extensive as the male's (top).

The Hooded is another of the fly-catching warblers. Like a number of other parulines, it possesses a broad bill depressed at the base in addition to rictal bristles. Flitting into the air to snatch a flying insect, the Hooded invariably fans its showy tail feathers, displaying the large areas of white. Its habit of opening and shutting its tail reminds one of the American Redstart's similar behavior.

Some field guides, in describing the female Hooded Warbler, note that she lacks the black hood. This is not altogether true. Young females may fit this description, but I have seen older females that have so much black in their plumage that they may easily be mistaken for males. At the nest above Dam no. 9 referred to earlier, the female and her mate were very similar. This misunderstanding may have led some early observers to their conclusion that the male shares incubation of the eggs with the female. We now know he does not.

Hooded Warbler is an obviously appropriate English name for this bird. The generic name, *Wilsonia* (will-SO-nih-ah), honors Alexander Wilson, the early American ornithologist; the species name, *citrina* (sit-RYE-nah), is from the Latin *citrus* or "lemon," referring to the bird's underparts.

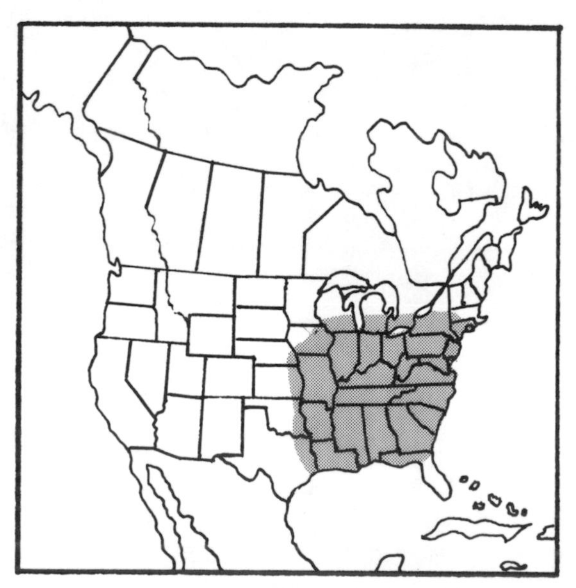

Breeding Range of Hooded Warbler

47. *Wilson's Warbler*

Wilsonia pusilla

PLATE 18

I have found three "colonies" of nesting Wilson's Warblers, two at the edges of Maine alder swales, one in a large willow thicket in Rocky Mountain National Park, Colorado. I am not sure that "colonies" is the correct description, for they were not colonial in the sense that seabirds, swallows, herons, and pelicans are. In areas where I studied and photographed the birds, each pair had its own territory which was well separated from a neighboring pair. Each male patrolled his territory and vigorously chased intruding Wilson's Warblers.

The feature that made each area so attractive that many pairs formed a "colony" was ideal habitat—the proximity of alder swales, the nearness of flooded willow bottomland, and the general mossy, boggy environment that offered choice nesting sites.

Most Wilson's Warblers nest on the ground, often at the base of a bush or sapling. The nest is well concealed in moss or in a grassy hummock. It is a bulky structure, large for such a small bird, and is built almost entirely of grasses with a few leaves or animal hairs added.

In Maine, near the town of Ellsworth, Hancock County, I observed a female Wilson's Warbler just starting her nest under a blackberry bush on a slope above an alder swale. She laid four eggs which she incubated for eleven days. Incubation began the evening before she laid her fourth egg. All four youngsters survived and left the nest exactly twenty-seven days after I watched the female carry the first material to the site.

At this nest in Maine, both the male and female wore a round black cap. I have found this typical of eastern Wilson's Warblers, though field guide pictures show capless females. In the Colorado colony a

An alder swale at the side of U.S. Route 1 near Ellsworth, Maine, was home to a "colony" of Wilson's Warblers. Nests were at the edge of the swale, not in it.

number of nesting females showed no sign of black on their heads. After reviewing the specimens in the Carnegie Museum of Natural History, Dr. Parkes reports: "Specimens . . . indicate that about ⅓ of spring and summer females from the west have black caps, about ⅓ have scattered black feathers or dusky caps, and about ⅓ have caps the same color as the back, as shown in the field guides. It is possible that those with no black at all are all first-year females, but this needs further study."[1]

None of the Wilson's Warblers' nests that I have found have contained cowbird eggs. Friedmann declares that Wilson's is "a fairly regular victim." Most records are for western races. Each of my Colorado nests had five warbler eggs; those in Maine held mostly clutches of four.

The Wilson's Warbler's song is loud and clear, a single note repeated rapidly, dropping downward in pitch toward the end. Peterson interprets it as *chi chi chi chi chi chet chet.* Others suggest *zee-zee-*

[1] Personal correspondence.

Most Wilson's Warblers nest on the ground, often at the base of a bush or sapling. The bulky structure is well hidden in moss or in a tussock of grass.

Female Wilson's Warblers may or may not have a black cap similar to the male's. Since the male is known not to incubate or brood, the bird here is almost certainly a female.

zee-zee-zee-e. In the West, the song is often mistaken for the similar song of the Orange-crowned Warbler; in the East, it is sometimes compared to the song of the Nashville Warbler, but it is less flowing, more abrupt.

In a study of the breeding ecology of Wilson's Warbler in the Sierra Nevadas in California, Stewart, Henderson, and Darling found polygyny practiced by 13 percent of the males. In a coastal area included in their studies, the males were monogamous, although the number of males in comparison to the number of females was the same in both areas. Nesting success in the Sierra population was determined to be 71 percent, considerably higher than the 33 percent in the coastal populations.

Nests of the Sierra birds were on or in the ground, while most of those of the coastal pairs were about eighteen inches to two feet above ground in blackberry vines or ferns. There were also other differences in nesting behavior. The authors suggest "that certain factors are advantageous to the development of polygyny, including paucity of

In autumn, this young Wilson's Warbler will follow the route of its parents to their wintering range in southern Texas and Louisiana south to Mexico and Central America.

breeding habitat, patchiness in habitat quality, and low predation pressure."[2] Studies of other species in varying habitats might also show similar differences in nesting strategies used by a single species.

In winter, Wilson's Warblers range from southern Texas and Louisiana to southern Mexico, the highlands of Central America, and Panama. Alexander F. Skutch writes that the Wilson's Warblers that winter in Guatemala are away from Central America for a little more than three months a year.

The species was named by Elliott Coues, in honor of Alexander Wilson, who had described it early in the nineteenth century. He called it Wilson's Black-capped Fly-catching Warbler. The name Black-capped Warbler is still used in some areas and in some publications. The western subspecies were formerly called Pileolated Warblers, referring to the black pileum, or crown. *Wilsonia* is the Latinized form of Wilson's name. The specific name *pusilla* (pew-SILL-ah) means "very small."

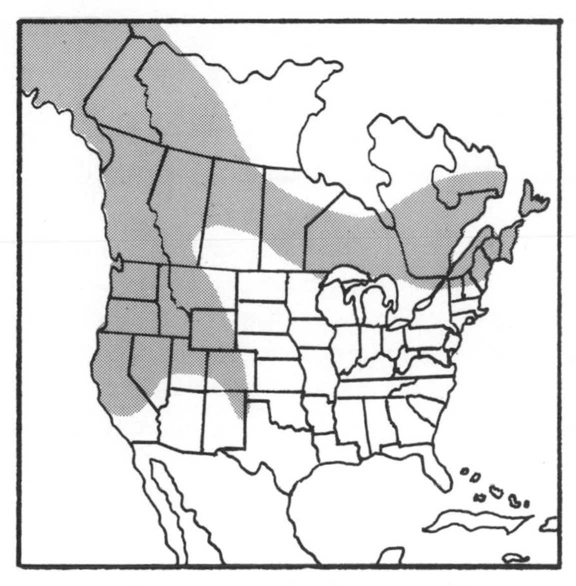

Breeding Range of Wilson's Warbler

[2] Stewart, Robert M., R. Phillip Henderson, and Kate Darling, "Breeding Ecology of the Wilson's Warbler in the High Sierra Nevada, California," *Living Bird,* 16th Annual (1977), p. 100.

48. Canada Warbler

Wilsonia canadensis

PLATE 23

At the nest of a Canada Warbler in Cook Forest State Park in Pennsylvania, I saw a male exhibit an interesting bit of instinctive behavior known to ornithologists as "anticipatory feeding." For several days the female had been sitting on four eggs. During a brief period when she had left the nest to feed, her mate approached it with food in his mouth. Landing on the rim of the nest, he appeared to offer the food to the eggs. After perhaps ten seconds of tendering the food, he ate it himself and flew away.

The same behavior was observed nine times by Herbert Krause at a Canada Warbler's nest in Cheboygan County, Michigan. During a total of thirty-nine hours Krause spent in a blind observing the Canada's nest during incubation, the male offered food to the eggs during the female's absence and each time ate the food himself. On the ninth occasion the female was on the nest when the male approached. This time he offered the food to her. At first she refused, but eventually she accepted it.

Krause theorizes that "the male's attempt to feed the eggs may have been a kind of 'anticipatory feeding' of the young." He further suggests, "Perhaps there is a kind of anticipatory building up of tension, an innate drive to feed the young which grows in intensity during the incubation period."[1] In other words, this behavior may amount to practice sessions before the male assumes his eventual responsibility for feeding his offspring. It makes me wonder: On the

[1] Krause, Herbert, "Nesting of a Pair of Canada Warblers," *Living Bird*, 4th Annual (1965), p. 10.

Canada Warblers are wilderness birds. They seek heavy woodland undergrowth, swamps, stream banks, rhododendron thickets, deep ravines, and moist brushlands.

many occasions when I have watched a male bird bring food to the nest while the female incubated, did he intend to feed her or was he following an instinctive pattern for feeding young?

Canada Warblers are wilderness birds. They seek heavy woodland undergrowth, swamps, stream banks, rhododendron thickets, deep ravines, and moist brushlands. During the breeding season, the Canada Warbler is associated with a wet habitat, but the nests I have found are typically near such habitat rather than in it. My Maine nests were buried in mossy hummocks, much like nests of the Yellow-bellied Flycatcher. In Pennsylvania, nests I have discovered were in the banks of streams, similar to nests of Louisiana Waterthrushes, or in rotted stumps.

No warbler with which I am acquainted protests more than the Canada when an intruder enters its territory. Usually, a chipping male warbler indicates a nest nearby, but this rule does not always hold

No warbler with which I am acquainted protests more than the Canada when an intruder enters its territory. On one occasion, a male chipped at me ninety-six times in one minute.

During the nesting season, Canada Warblers are associated with wet habitat, but nests are likely to be near such habitat rather than in it.

true for the Canada. Onc pair in Maine resented my presence as soon as I was within two hundred yards of their nest. There seems to be survival value in such behavior. I have searched long and hard for a Canada Warbler's nest with both birds protesting beside me only to find the nest later some distance away.

On one occasion while I was searching for a nest, the male's incessant chipping was so loud and annoying that I started to count his protests. For several minutes, he averaged ninety-four chips per minute. So constant were his scolds that ninety-two was the fewest times he chipped in any one minute and ninety-six was the most.

I discovered still another bit of unusual behavior when I set up my camera equipment to photograph Canada Warblers feeding nestlings. I settled myself fifty feet away from the nest with a remote-control apparatus beside me. The female paid no attention at all to the camera within two feet of her nest. But she spent several hours fluttering at my feet, dragging her wings and trying to lure me away. On the other hand, the male paid no attention to me or the camera. He continued to feed the young regularly during the entire photographic session.

There is no satisfactory way to interpret the Canada Warbler's song, and there seems to be no other warbler's song with which the Canada's can be compared. One helpful fact is that the Canada's song lacks the buzzy, trilling or husky notes of many warblers. The song consists of from two or three notes to as many as eight, delivered as a jumbled, disconnected warble ending with an emphatic *wip.* Two notes in succession are rarely on the same pitch. Several writers compare the quality of the notes to that of the Common Yellow-throat. Francis H. Allen, in Bent's *Life Histories of North American Wood Warblers,* wrote the song as *te-widdle-te-widdle, te-widdle-te-wip.* Kenneth C. Parkes finds it much like the song of the Magnolia Warbler, but decidedly longer.

The Canada Warbler winters in South America, principally in Colombia and in the mountains of Venezuela south to eastern Peru. In spring migration, it does not reach its breeding grounds until the first or second week of May. The fall movement begins very early, long before there is any apparent necessity for the birds to return to their winter haunts. By August, fall migration is well under way.

The Canada is closely related to the Hooded and Wilson's Warblers. The specific name *canadensis* is Latin for "of Canada." Although the bird does nest in Canada, it also breeds in the mountains as far south as Georgia. The species was first seen in Canada and described by a French ornithologist named Brisson. It was one of the birds introduced by Linnaeus in his twelfth and last edition of the *Systema.*

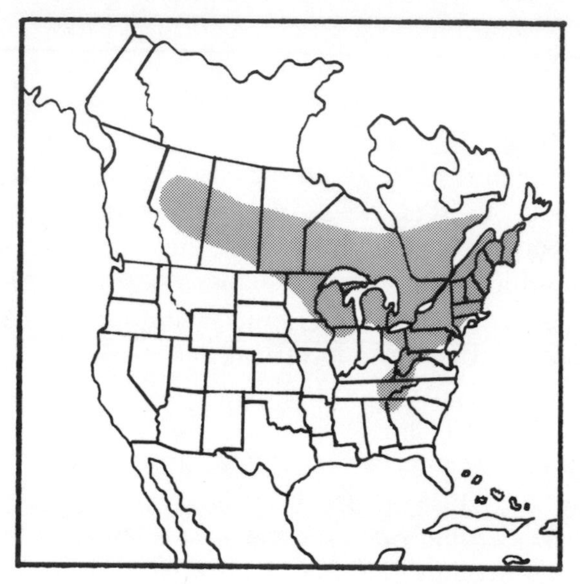

Breeding Range of Canada Warbler

49. *Red-faced Warbler*

Cardellina rubrifrons

PLATE 21

Long ago I learned that it is usually a mistake to declare that any species of warbler "always does this" or "never does that." It is surprising to learn, often through sad experience, that birds are individuals and that the behavior of one is not necessarily an indication of how another of the same species will act.

For example, my experiences with Red-faced Warblers at their nests were exactly the opposite of those reported by Herbert Brandt in his *Arizona and Its Bird Life.* Brandt stated that "in no case was the Red-faced Warbler helpful in disclosing its cradle." He added, "Furthermore, when watched, it refused to return to its abode, even when that was filled with hungry young."[1]

At four nests that I found in the Chiricahua and Catalina Mountains of southeastern Arizona, the birds were anything but shy. At one nest where the adults were feeding young and the female was brooding I took nineteen photographs in a few hours. Both birds accepted my camera on a tripod two feet from the nest. They completely ignored the flash that accompanied each exposure.

The Red-faced is one of the Wood Warblers that Dr. George M. Sutton called "the United States warblers which inhabit Mexico." Its principal breeding grounds are in the mountains of Sonora, Chihuahua, Sinaloa, and Durango in northern Mexico. The Red-faced Warblers that nest in Arizona and New Mexico are northern outliers of a Mexican species. Migration in late summer or early fall means moving

[1] Brandt, Herbert, *Arizona and Its Bird Life,* (Cleveland: The Bird Research Foundation, 1951), p. 496.

Red-faced Warblers nesting in Arizona and New Mexico are tropical birds that have extended their range northward. Migration, both spring and fall, is short.

High mountain forests of conifers and aspens are home to the Red-faced Warbler in summer. Here at an elevation of sixty-four hundred feet and upward, the birds nest on the ground.

southward a distance that is very short compared to journeys made by many other parulines. The birds move north from their winter home in March.

The Red-faced is a bird of high mountain forests and nests at an elevation of sixty-four hundred feet and upward, where there are conifers and aspens. The nest, however, is on the ground, not in the towering pines, Douglas firs, and Englemann spruces that shade it. It is commonly in a litter of fallen leaves and buried so deeply that it may be completely out of sight. Nests I have found have all been on mountainsides or canyon slopes.

As in the case of most ground-nesting birds, the nest of the Red-faced Warbler is difficult to find when the female is incubating. She is a tight sitter and will flush only when approached closely. When caught building a nest or feeding young, the birds have not seemed unduly alarmed when I observed them.

I agree with those observers who liken the song of the Red-faced Warbler to that of the Yellow Warbler. Dr. Walter Penn Taylor, in

The Red-faced Warbler builds its nest in a litter of leaves and low vegetation. It may be buried so deeply that it is completely out of sight.

Bent's *Life Histories of North American Wood Warblers,* calls it a whistled song with variations, *a tink a tink a tink twee teee tswee tsweep.* He says it is more ringing and bell-like than the song of Grace's Warbler, which is often associated with it in the pines.

In addition to gleaning among the branches of conifers, the Red-faced Warbler indulges in considerable fly-catching. While striking out for passing insects, this beautiful bird shows most conspicuously one of its field marks, the white rump patch.

Full-grown juveniles are very dull-looking, with no trace of red, but before the end of August they pass through a molt and acquire the brilliant color pattern of their parents.

The Red-faced Warbler's English name is apt, and the bird is so different from all other warblers that it is unmistakable. The species name, *rubrifrons* (RUBE-rih-fronz), takes note of its appearance; it is Latin for "red-fronted." *Cardellina* (car-dell-INE-ah) is Latin for a kind of finch. The Red-faced Warbler was first described in 1841 by an American named Giraud, about whom little has been written.

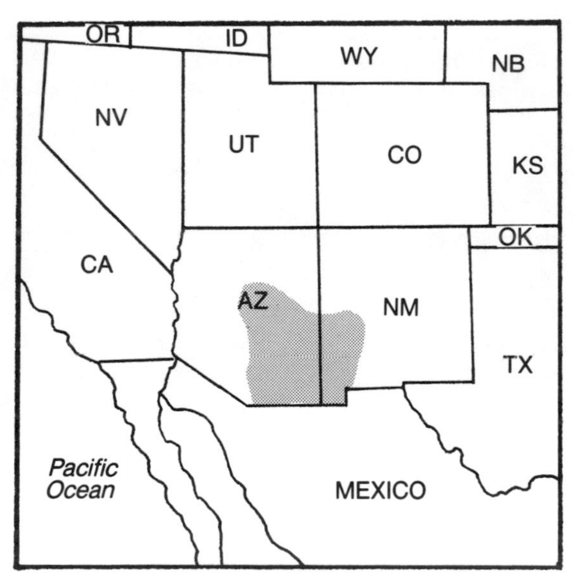

Breeding Range of Red-faced Warbler

50. Painted Redstart

Myioborus pictus

PLATE 21

During the early years of my career as a wildlife photographer I looked forward to the day when I would visit the mountains of southeastern Arizona and photograph the Painted Redstart. Paintings I admired all showed the bird to best advantage: the brilliant scarlet breast and belly accentuated against back and wings of velvety black, broken only by a wide white wing patch. I knew that the Painted Redstart nested on the ground and I knew that the sexes were alike, so I expected it to be easy to acquire the photographs I wanted.

The day came when I focused my camera on a nest with young, but recording that bright red belly was not nearly as easy as I had anticipated. Both adults cooperated, coming regularly to feed the nestlings; but they always perched with their black backs to the camera. My next step was to move the camera to the side where the red front would be visible. They weren't fooled; they came to a different spot on the nest rim where their backs were again toward the camera. I could see no red feathers. At last, when I caught a bird approaching the nest from above, I finally recorded the picture I wanted. But whoever said that bird photography was easy!

For many years the Painted Redstart was considered a close relative of the American Redstart; they were placed in the same genus, *Setophaga* (see-TOFF-ah-gah), by William Swainson, who first saw the bird in Mexico in about 1829. But recently taxonomists have declared the Painted Redstart more closely related to a group of Central American Wood Warblers in the genus *Myioborus* (my-yoh-BORE-us). This name is formed from two Greek words, *myia,* "a fly," and *borus,* "greedy," and probably refers to its fly-catching feeding habits. The specific name, *pictus,* is Latin for "painted."

The Painted and American Redstarts once were considered close relatives. Changes have placed the Painted Redstart, above, in a different genus. It is a tropical bird with a limited range in the United States.

All thirteen Painted Redstart nests I have found have been in mountains, on the ground in steep locations. These nests were on a bank of a narrow canyon, a steep cliff, a canyon wall, the top of a cliff, or the side of a dry wash. Grass hiding this nest had to be pushed aside to photograph it.

Of ten tropical redstarts of the genus *Myioborus,* the Painted is the only species that breeds in the United States. Its range is limited mainly to the high mountains of southeastern Arizona and southwestern New Mexico, with a few nesting records for the Chisos Mountains of southern Texas, and one record for the Laguna Mountains in California in 1974. It is really a Mexican and Central American species, breeding as far south as Nicaragua. The limited numbers that nest within our borders are typically migratory, although a few individuals may occasionally be found wintering in southern Arizona and California. Breeding birds return to these more northern mountain homes in late March or early April and remain until family groups gather for postbreeding wanderings before heading for their winter range in early fall. In the late 1970s the Slate-throated Redstart, *Myioborus miniatus,* was seen in Arizona on several occasions, but so far there has been no evidence of nesting.

All thirteen of the Painted Redstart nests that I saw in the Santa Rita and Chiricahua Mountains were in steep locations. My notes read: "bank of a narrow canyon, rocky bank, steep cliff, canyon wall, top of a cliff, on a rock, side of a dry wash." Of the nests I found, only one was parasitized by the Brown-headed Cowbird. This nest, under a

Cave Creek Canyon in the Chiricahua Mountains of southeastern Arizona is ideal habitat for Painted Redstarts.

clump of grass and hidden by a blooming columbine overhead, held three warbler and two cowbird eggs.

Judy Marshall and Russell P. Balda, in their study reported in "The Breeding Ecology of the Painted Redstart," determined that polygyny is normal in this species, at least in a segment of the population. On one occasion, a banded male was seen feeding banded young that had fledged from two different nests. He showed no preference for young from either nest.

Unlike most Wood Warblers, the female Painted Redstart sings, sometimes dueting with her mate during courtship. The primary song is composed of seven loud and ringing syllables and is commonly interpreted as *weeta weeta weeta wee.* H. S. Swarth described the call note as "like the peep of a young chicken." To my ears, the call note was a loud and distinct *cheap,* similar to the alarm note of a House Sparrow.

This species is unlike most other Wood Warblers in that young Painted Redstarts acquire a nearly adult plumage at the postjuvenal molt. After that molt, sexes are indistinguishable in any plumage throughout the year. This is one of the characteristics of the genus *Myioborus,* and was one of the many factors influencing the change in classification.

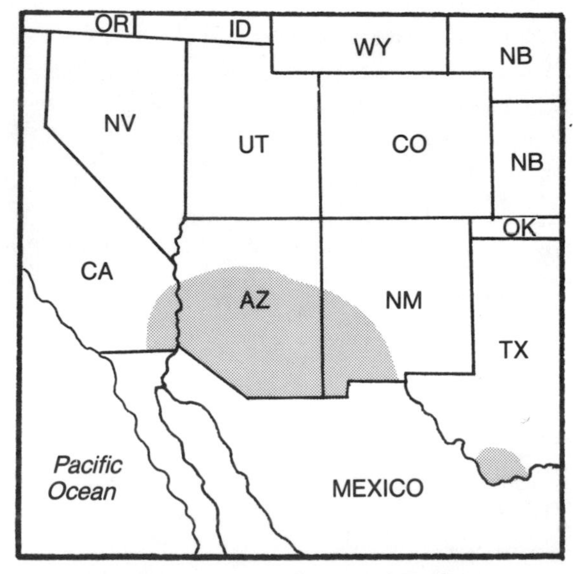

Breeding Range of Painted Redstart

51. *Rufous-capped Warbler*

Basileuterus rufifrons

NO PLATE

The basis for including the Rufous-capped Warbler in a list of Wood Warblers that nest within the United States and Canada is the discovery of a nest with four eggs in Cave Creek Canyon, Chiricahua Mountains, Cochise County, Arizona, on July 19, 1977. The apparent owner of this nest suddenly disappeared four days later, leaving behind four unhatched eggs and a question regarding their origin. Only one bird was seen near the nest; and since sexes are indistinguishable in the Rufous-capped Warbler it was never certain whether this was an unmated female who built a nest and laid infertile eggs. In Cave Creek Canyon on May 9, 1977, a Rufous-capped Warbler had been seen in the state for the first time. This individual vanished promptly, but the nest found in July was only about 1.3 miles from where the bird was seen in May.

On April 18, 1978, a Rufous-capped Warbler was seen and studied in Cave Creek Canyon less than a half-mile from the previous year's unsuccessful nest. This bird was not seen again after April 18.

The most recent reported sight of this species in the United States was turned in by participants in the Boerne, Texas, Audubon Christmas Count on January 3, 1982. In the 1970s there were three Texas records, one at Falcon Dam and two in the Big Bend area. One of these Big Bend birds was seen several times in Santa Elena Canyon between March and June. It was associating with several Yellow-breasted Chats, and there was speculation that the Rufous-capped tried to mate with one of the Chats. It has been suggested that the Arizona and Texas sightings are plausible indications of a northern

One nest of the Rufous-capped Warbler discovered in Cave Creek Canyon, Cochise County, Arizona, is the basis for including this tropical warbler among those that nest in the United States. The bird laid four eggs, all infertile.

extension of the bird's breeding range. This is the northernmost species of a very large genus found throughout the New World tropics.

This species, like other members of its genus, builds a roofed, arched, or oven-shaped nest on the ground. Ordinarily these warblers nest in southern Mexico, British Honduras, and central Guatemala. Typical habitat includes open woodlands and brushy hillsides.

The name *Basileuterus* (bas-i-LOO-ter-us) is derived from a Greek word *basileus,* "a king." The species name is formed from Latin words, *rufus* meaning "red" and *frons* meaning "forehead, brow."

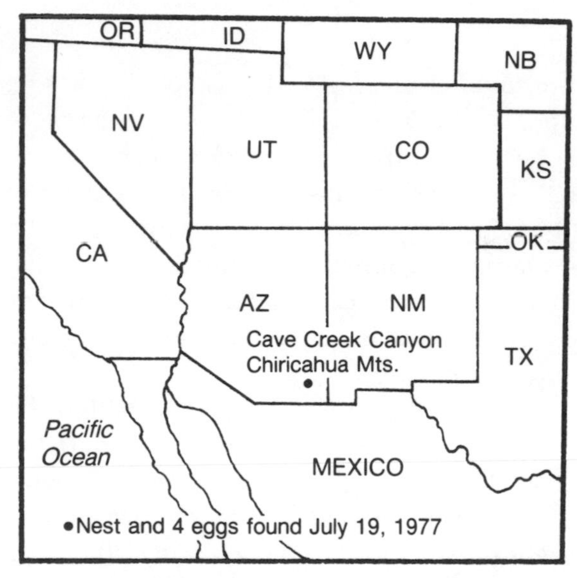

Breeding Range of Rufous-capped Warbler

52. *Yellow-breasted Chat*

Icteria virens

PLATE 23

Since 1901 the status of the Yellow-breasted Chat has been in doubt. At that time in *The Warblers of New England* Charles Johnson Maynard, a Massachusetts naturalist, wrote the following:

"Just how this bird resembles the Warblers, except in having nine primaries, it is difficult to state. The differences are, however, quite obvious. It is a large bird, about twice the size of the largest Warbler. The bill is thick, with no indication of a notch at the top; the tarsi are without scales. The tongue is not cleft at the top nor ciliated, and the stomach is quite muscular."[1]

Maynard might have added other ways in which the Chat differs from most other warblers. It lacks natal down and has a complete postjuvenal molt. In breeding condition, the male Chat has a black mouth lining; the female's mouth lining is red or pink. The nest and eggs show very little similarity to those of any other warbler.

Most obvious, however, is its unwarblerlike behavior. Its habit of holding food with its foot is unknown in other warblers. The aerial antics of this species during courtship displays are unlike those not only of any other warbler but of any other bird. Its song is unique; it sometimes sings at night and sometimes seems to mimic the songs of other birds.

Dr. Kenneth C. Parkes was kind enough to explain to me the present position of the AOU regarding the status of the Yellow-breasted Chat. In a letter to me he wrote:

[1] Maynard, Charles Johnson, *The Warblers of New England* (West Newton, Massachusetts: C. J. Maynard, 1901).

Not all ornithologists believe that the Yellow-breasted Chat is a warbler, a member of the paruline subfamily. Its behavior is most unwarblerlike.

"Although the Chat has many unwarblerlike characteristics, nobody has ever been able to suggest a plausible relationship to any other group of birds. Charles Sibley's work with DNA-hybridization techniques indicates that the Chat is, indeed, a Wood Warbler, with no genetic affiliation with any non-warbler group. It may be a distinct offshoot, with no close relatives, from the ancestral warbler line, but IS, nevertheless, to be considered a paruline."

I have heard the Yellow-breasted Chat sing many times at night in Pennsylvania's overgrown pastures and hawthorn thickets. It is not something that one might describe as an occasional burst of song; rather it is a vigorous, wide-awake intentional medley of odd noises that may continue for long periods of time. To describe the bird's whistles, gurgles, chortles, catcalls, and grunts is almost impossible. Nevertheless, like so many other observers, I succumbed to the temptation to put into words what this clownlike bird says when it goes on one of its singing rampages: the alarm call of a wren; a series of nasal quacks; a wolf whistle; a foghorn; and a chuckling, high-pitched laugh. If you can combine all that into one voice, you have the love song of the Yellow-breasted Chat.

Usually this crazy song is given as the bird remains well concealed; but on other occasions the bird flies into the air, and then drops slowly with its wings flapping over its back, its tail jerking up and down, and its legs dangling loosely at full length, all the while uttering a long series of indescribable notes. Many describe the Chat as a ventriloquist, for it seems first in one place and quickly in another. The alarm note is an emphatic *chee-uck.*

If it were not for the bird's outlandish behavior, the Yellow-breasted Chat might often be overlooked, for it frequents the densest thickets and is very suspicious and alert when a stranger enters its domain.

Despite the fact that the Chat's nest is rather bulky for a warbler, it is not easy to find. The bird usually selects a place two to six feet above ground in a brier thicket, a hedgerow, or a hawthorn thicket to which access is very difficult.

Not unexpectedly, the Chat lays the largest eggs of any warbler that nests in the United States. It commonly lays four, but may lay three or five. They are similar to the eggs of the Ovenbird; however, the nests could not possibly be confused.

Yellow-breasted Chats nest in areas where access is difficult: brier thickets, hedgerows, and hawthorn thickets. The nest is bulky but not easy to find.

Nest and eggs of the Yellow-breasted Chat are different from those of other warblers. The nest is large, and the Chat lays the largest eggs of any warbler that nests in the United States.

Many investigators have commented upon the Chat's extremely timid nature. This has not been my experience, and I agree with George A. Petrides, who studied the life history of the species and stated: "The suspicious nature of this species is believed overemphasized in the literature."[2] I have photographed a number of Chats at their nests and have found them no shier than many other species of warblers and, in some cases, not as timid as others (e.g., Virginia's and Canada Warblers).

The Yellow-breasted Chat is said to be a common victim of the Brown-headed Cowbird, although I have not found it to be so. Of ten nests I studied in Pennsylvania in 1964, none was parasitized. In that same area, in adjacent woodlands, eleven of twelve nests of Wood

[2] Petrides, George A., "A Life History Study of the Yellow-breasted Chat," *Wilson Bulletin,* vol. 50 no. 3 (1938), p. 189.

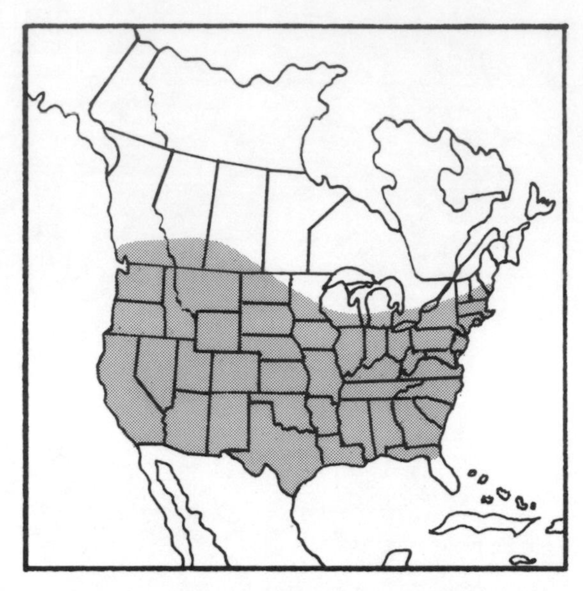

Breeding Range of Yellow-breasted Chat

Thrushes were parasitized. In Michigan, Walter P. Nickell found a Chat's nest with five eggs of the cowbird and two of the Chat. The Chat was on the nest and the eggs were warm.

Among the characteristics that distinguish the Yellow-breasted Chat from other members of the Parulinae is that of "wandering" northward after the breeding season. Records of Chats suddenly appearing in northern states, particularly New England, and also in the Maritime Provinces of Canada, are common in August, September, October, and occasionally into winter. From banding records it is evident that this tendency has increased in recent years.

Most Chats, except for the wanderers, migrate early. In the northern limits of its range, the Chat may be gone in August, and very few are left anywhere in its breeding range by early September. Toward the end of September they begin to arrive in their winter home in Mexico, Central America, and Panama. Only rarely do they winter as far north as southern California, eastern Texas, and southern Florida.

The generic name of the Chat, *Icteria* (ick-TER-ih-ah) is from the Greek *iktéros,* meaning "jaundice," hence yellow; the species name, *virens* (VIR-enz), is Latin for "green," referring to the olive green back. In England, the word *chat* is used as a suffix for bird names, as in *stonechat* and *whinchat.* It has been suggested that the American warbler may have reminded early colonists of some of the vocal notes of the Old World birds, thus suggesting its name. The Yellow-breasted Chat was one of the first birds recorded for the American colonies. Mark Catesby, the English author and illustrator, considered to be the first of America's great naturalists, included the Yellow-breasted Chat in his book *Natural History of the Carolinas,* published in about 1731.

53. Olive Warbler

Peucedramus taeniatus

PLATE 22

The day may come when the Olive Warbler no longer belongs to the Wood Warblers' world because the AOU Check-list Committee will have concluded that it is *not* a Wood Warbler. Many taxonomists have already recommended that this action be taken, as the Olive Warbler bears several features not characteristic of any other members of the paruline family. There are differences of opinion as to just how the bird should be classified, but there seems to be some agreement that it is similar to the Old World Sylviinae, a group represented in our country by the kinglets and gnatcatchers.

That the Olive is somewhat of an oddball among warblers came to my attention when I had the opportunity to study two pairs at their nests in Barfoot Park, eight thousand feet up in the Chiricahua Mountains of southeastern Arizona. It was impossible for me to photograph either nest. Both were in ponderosa pines, one fifty feet above ground and twenty feet from the trunk, the other forty-five feet up and fifteen feet out.

I was fortunate enough to have an adult male at one nest, and, at the other, an immature male in first-year plumage, a bird quite unlike his fully adult counterpart. The male Olive does not acquire adult plumage until after its first breeding season. He is, however, capable of mating in this subdued dress.

As far as I could determine, only two pairs of Olives were nesting in Barfoot Park that June; and that meant that the birds had almost unlimited territory for roaming and foraging. The males' loudly whistled *peedo peedo peedo peedo* came to me from high in the pines throughout the day. Although there are slight variations (*peter peter*

peter peter and *tiddle tiddle tiddle ter*), the basic song is very similar to the song of the Tufted Titmouse.

Like Grace's Warbler, which also nests in Barfoot Park, the Olive Warbler has a characteristic way of approaching the nest. Instead of flying directly to it, the birds land near the trunk and quietly creep along the branch to the nest hidden by clusters of pine needles near the tip.

I was unable to examine the nests closely; but William G. George, who collected eight nests, declared that none resembled any other Wood Warbler nests that he had seen. Among other materials used in construction by the female are soft white plant fibers peeled from the under surface of living leaves of the silver-leafed oak. Wirelike root-

High in the ponderosa and Apache pines in the mountains of southeastern Arizona and southwestern New Mexico, the Olive Warbler, a tropical species, makes its summer home. Its status as a Wood Warbler has been questioned.

Olive Warbler eggs are unique. Almost the entire shell is covered with a smudge that makes the surface appear nearly black.

lets form a circular framework into which the female piles the white plant fibers for the inner wall of the nest.

Olive Warbler eggs are unique. Herbert Brandt described them this way: "I beheld four of the strangest warbler eggs in all my experience; scarcely could I believe my eyes. Almost the entire shell of each was covered with a dark smudge of markings which made the surface appear nearly black. Then suddenly I recalled that Frank Willard had once remarked to me: 'The Olive Warbler should be called the Black-egged Warbler.' "[1]

Another unwarblerlike characteristic of the Olive Warbler is described by George: "Nests of Peucedramus in which young have been successfully reared are found to be soiled with excrement (three specimens), as are the needles immediately adjacent to the nests. The soiling seems to occur during the final days that the young spend in or about the nest. The great majority of droppings are deposited on

[1] Brandt, p. 563.

the nest rim and sides, almost none in the cup."[2] George points out that he has never known of another report of this trait in Wood Warblers. (See similar observation in chapter 9, "Lucy's Warbler.")

The Olive is primarily a species of the mountains in Mexico and Central America that barely reaches the southwestern United States. Olive Warblers are resident birds in most of their range, including a few as far north as the Santa Catalina Mountains in Arizona. Most of the more northern breeders, however, retire in winter to Mexico and south to Nicaragua.

The Olive's preference for pines influenced the naming of this bird. *Peucedramus* (pew-SED-rah-mus) is from the Greek *peuke* for pine and *dramein* meaning to run, thus "a runner in pines"; *taeniatus* is Latin for banded, referring to the black mask through the eyes of the males. The species was first recorded in the United States when Henry W. Henshaw captured three specimens in Arizona in 1874, but tracing the history of how the Olive Warbler acquired its English and Latin names has turned out to be as difficult as photographing its nest.

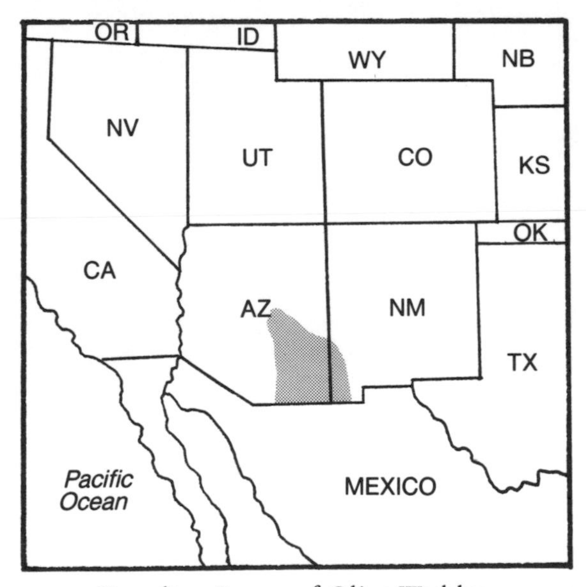

Breeding Range of Olive Warbler

[2] George, William G., *The Classification of the Olive Warbler* (*Peucedramus taeniatus*), American Museum Novitates, no. 2103 (New York: American Museum of Natural History, 1962), p. 37.

Glossary

Allopatric species: Species whose geographic ranges do not overlap.

Altricial young: Birds that are helpless when they hatch.

AOU: American Ornithologists' Union. An organization whose purpose is advancing the science of ornithology. A committee of its members compiles a checklist (spelled *Check-list*) giving the classification, names (both English and scientific), and ranges of North American birds, which is generally accepted as the authority.

Arboreal: Tree-inhabiting.

Banding: The placing of individually numbered metal bands (and sometimes additional colored bands) on the legs of birds in order to identify individuals and to gather data on migration, behavior, longevity, population changes, return to nesting sites, and more. In Great Britain and Europe, the process is called ringing.

Biome: A major community, such as a coniferous forest, deciduous forest, tundra, grassland, or swamp, combined with the animal life that inhabits it. Biomes are also called *biotic communities.*

Brood: The number of birds resulting from a single nesting.

Brooding: The providing of warmth and shelter to nestlings by a parent bird.

Casual: Said of birds that occur occasionally, but without regularity, in a given area.

Clutch: A complete set of eggs laid by one female in one nesting.

Colonial: Used to describe birds such as terns, herons, gulls, pelicans, seabirds, and others that build their nests close together in groups.

Coniferous: Referring to cone-bearing trees or shrubs such as pines, spruces, firs, hemlocks, and others.

Deciduous: Referring to trees or other plants that shed their leaves at the end of the growing season.

Ecology: The study of the relationship between birds (or other organisms) and their environment.

Endangered species: Plants and animals that are in danger of extinction throughout all or a significant portion of their range, as determined by the Fish and Wildlife Service of the United States Department of the Interior. Bachman's and Kirtland's Warblers are the only Wood Warblers on the list.

Epiphyte: A plant, such as Spanish moss, that grows upon another plant, using it merely as a support, not as a source of nourishment.

Family: In terms of the scientific classification of birds (or other organisms), a group of closely related genera.

Fecal sac: A small, whitish sac or membrane containing excrement and partly digested food of nestling birds. Among Wood Warblers, parents either eat them or carry them away.

Fledgling: A young bird that has left the nest but is still dependent upon its parents for food.

Flush: To rouse and cause to start up suddenly or fly away.

Forb: A herbaceous plant that is not a grass or grasslike.

Gape: A bird's open mouth.

Genus (plural, *genera*): A group of related species descended from a common ancestor. The name of the genus is the first part of the scientific name for species in the genus.

Gregarious: Occurring in flocks or groups.

Habitat: The specific environment in which a bird lives.

Hummock: A small, low, rounded hill, mound, or knoll higher than the surrounding area; often a mossy knoll in a swamp or marsh.

Hybrid: A bird produced by the cross-breeding of different species.

Immature: As used in this book, not having yet acquired complete adult plumage.

Incubation: The action of a bird sitting on its eggs to apply heat to them.

Incubation patch: A featherless area of bare skin on the belly or underpart of a bird's body that, during the breeding season, is thickened and develops a rich supply of blood vessels under the skin. By pressing this patch against the eggs, the bird transfers heat to them from its body.

Incubation period: The time that elapses between laying of the last egg in a clutch and the hatching of that egg.

Indigenous: Native to a particular area.

Life list: Term used by bird watchers indicating the number and names of the different species of birds seen during a lifetime of birding.

Lore: That part of each side of a bird's face between the eye and the bill.

Lumping: In taxonomy, combining groups (such as genera or species) formerly considered separate. "Lumpers" emphasize *similarities,* hence tend to combine related birds into larger groups. See **Splitting**.

Mist netting: Catching birds for scientific purposes, usually banding, by the use of mist nets, so named because the fine nylon (or other thread) of the mesh is almost invisible against the background of the birds' habitat.

Molt: Periodic replacement of feathers, during which old feathers are cast off and replaced by new ones.

Monogamy: A mating system in which only one male and one female mate.

Morphology: The branch of biology dealing with the physical form and structure, both external and internal of plants and animals.

Nestling: A young bird in a nest during the period between hatching and fledging.

Nomenclature: A system of names or terms used in a particular science, such as ornithology. The scientific names used in ornithology are determined by the International Code of Zoological Nomenclature.

Parasitism: A situation in which an organism depends on another for its existence while offering nothing in return. The Brown-headed Cowbird depends on other birds to incubate its eggs and to feed its young. The host birds appear to gain no advantage.

Passerine: A species belonging to the order Passeriformes, the largest order of birds, comprising more than half of all species. Loosely speaking, passerines have evolved more recently than other groups of birds, have a voice box and special muscles that can produce singing, and have special feet and leg muscles which enable them to grasp a perch.

Polygyny: A form of polygamy in which a male forms pair bonds with more than one female in a single season. In such matings, females assume most or all of the parental duties.

Postnuptial: Occurring after the breeding season.

Primaries: The major flight feathers of the wings of birds.

Race: See **Subspecies**.

Rictal bristles: Stiff, hairlike feathers surrounding the mouth of a bird, effectively enlarging its gape.

Scree: Steep mass of small rocks on the side of a mountain.

Sedentary: Nonmigratory.

Slashings: An area strewn with the debris of trees left after lumbering or from wind or fire.

Songbird: A species belonging to the suborder Oscines of the order Passeriformes.

Speciation: The evolutionary process whereby a population of birds that were formerly of one species has been divided into two or more species.

Splitting: In taxonomy, subdividing groups (such as genera or species). "Splitters" emphasize *differences,* hence tend to separate birds into smaller groups. See **Lumping**.

Stub: A dead tree, or dead part of a live tree, that provides nest sites well above ground for cavity-nesting birds such as the Prothonotary Warbler.

Stump: That part of a tree which remains after lumbering or after a tree has broken off near the ground.

Subspecies: A local population or race of a bird with characteristics visibly different from other populations of the same species, but which can still interbreed with those populations.

Swale: A low wet, swamplike area surrounded by higher ground; usually brushy, and often choked with vegetation such as alders.

Sympatric species: Separate distinct species whose distribution ranges overlap.

Talus: A sloping mass of rocky fragments at the base of a cliff.

Taxonomy: The science of classifying, describing, and naming organisms, including birds, other animals, and plants. It includes grouping birds into orders, families, genera, species, and subspecies.

Territory: An area that a bird, typically a male, defends against intrusion by other birds of the same species during the breeding season. Some birds defend feeding areas also, in both the breeding and wintering ranges.

Type locality: The place where a type specimen was collected.

Type specimen: The individual bird specimen on which the original scientific description of a species, or subspecies, was based.

Bibliography

Abbott, Clinton G. "Peculiar Nesting Site of a Dusky Warbler." *Condor,* vol. 28 (1926), pp. 57–60.

Allen, Arthur A. *Stalking Birds with a Color Camera.* Washington, D.C.: The National Geographic Society, 1951.

American Ornithologists' Union. *Check-List of North American Birds.* 5th ed. Baltimore, Maryland: American Ornithologists' Union, 1957.

American Ornithologists' Union. *Check-List of North American Birds.* 6th ed. Washington, D.C.: American Ornithologists' Union, 1983.

Anderson, Walter L., and Robert W. Storer. "Factors Influencing Kirtland's Warbler Nesting Success." *Jack-Pine Warbler,* vol. 54, no. 3 (1976), pp. 105–115.

Arbib, Robert and Tony Soper. *The Hungry Bird Book.* New York: Taplinger Publishing Company, 1971.

Axtell, Harold H. "The Song of Kirtland's Warbler." *The Auk,* vol. 55, no. 3 (1938), pp. 481–491.

Bagg, Aaron Clark and Samuel Atkins Eliot, Jr. *Birds of the Connecticut Valley in Massachusetts.* Northampton, Massachusetts: The Hampshire Bookshop, 1937.

Bailey, Alfred M., and Robert J. Niedrach. *Birds of Colorado,* vol. 2. Denver: Denver Museum of Natural History, 1965.

———. "Nesting of Virginia's Warbler." *The Auk,* vol. 55, no. 2 (1938), pp. 176–178.

Bailey, Florence Merriam. *Birds of New Mexico.* Santa Fe: New Mexico Department of Game and Fish, 1928.

———. *Birds Recorded from the Santa Rita Mountains in Southern Arizona,* Pacific Coast Avifauna no. 15. Berkeley, California: Cooper Ornithological Club, 1923.

———. *Handbook of Birds of the Western United States.* Boston: Houghton Mifflin Company, 1902.

Baker, Bernard W. "Nesting of the American Redstart." *Wilson Bulletin,* vol. 56, no. 2 (1944), pp. 83–90.

Baker, David E. "Tennessee Warbler Nesting in Chippewa County, Michigan." *Jack-Pine Warbler,* vol. 57, no. 1 (1979), p. 25.

Banks, Richard C., and James Baird. "A New Hybrid Combination." *Wilson Bulletin,* vol. 90, no. 1 (1978), pp. 143–144.

Bankwitz, Kenneth G., and William L. Thompson. "Song Characteristics of the Yellow Warbler." *Wilson Bulletin,* vol. 91, no. 4 (1979), pp. 533–550.

Barlow, Chester. "Nesting of the Hermit Warbler in the Sierra Nevada Mountains, California." *The Auk,* vol. 16 (1899), pp. 156–161.

Bedayn, Barbara. "The Golden Wilson's Warbler." *Bird Watcher's Digest,* vol. 2, no. 4 (1980), pp. 22–23.

Bent, Arthur Cleveland, ed. *Life Histories of North American Wood Warblers.* Washington, D.C.: U.S. National Museum Bulletin no. 203, 1953.

Bigglestone, Harry C. "A Study of the Nesting Behaviour of the Yellow Warbler." *Wilson Bulletin,* vol. 25, no. 1 (1913), pp. 49–67.

Blake, Emmet R. "The Nest of the Colima Warbler in Texas." *Wilson Bulletin,* vol. 61, no. 2 (1949), pp. 65–67.

Bond, James. *Birds of the West Indies.* 2d ed. London: Collins, 1971.

———. "The Blackpoll Warbler in Maine and the Maritime Provinces." *Bulletin Maine Audubon Society,* vol. 7, no. 1 (1951), pp. 2–7.

———. *Native Birds of Mount Desert Island and Acadia National Park,* 2d rev. ed. Philadelphia: The Academy of Natural Sciences of Philadelphia, 1971.

Borror, Donald J. "Songs of the Yellowthroat." *Living Bird.* Sixth Annual. (1967), pp. 141–161.

Bowles, Charles Wilson. "Notes of the Black-throated Gray Warbler." *Condor,* vol. 4 no. 1 (1902), pp. 82–85.

Bowles, Charles Wilson, and John Hooper Bowles. "The Calaveras Warbler in Western Washington." *Condor,* vol. 8 (1906), pp. 68–69.

Bowles, John Hooper. "The Hermit Warbler in Washington." *Condor,* vol. 8 (1906), pp. 40–42.

———. "A Few Summer Birds of Lake Chelan, Washington." *Condor,* vol. 10 (1908), pp. 191–193.

Brand, Albert Rich. "Bird-song Study Problems." *Bird-Lore,* vol. 38 (1936), pp. 187–194.

Brandt, Herbert. *Arizona and Its Bird Life.* Cleveland: The Bird Research Foundation, 1951.

Brewster, William. "The Black-and-yellow Warbler (*Dendroica maculosa*)." *Bulletin Nuttall Ornithological Club,* vol. 2 (1877), pp. 1–7.

———. "Notes on Bachman's Warbler (*Helminthophila bachmanii*)." *The Auk,* vol. 8 (1891), pp. 149–157.

Brooks, Maurice. "The Breeding Warblers of the Central Allegheny Mountain Region." *Wilson Bulletin,* vol. 52 (1940), pp. 249–266.

———. "Cape May Warblers Destructive to Grapes." *The Auk,* vol. 50 (1933), pp. 122–123.

Brooks, Maurice, and William Clarence Legg. "Swainson's Warbler in Nicholas County, West Virginia." *The Auk,* vol. 59 (1942), pp. 76–86.

Brown, Richard D. "Yellowthroat Caught in Common Burdock." *Wilson Bulletin,* vol. 82, no. 4 (1970), pp. 464–465.

Brush, Alan H., and Ned K. Johnson. "The Evolution of Color Differences Between Nashville and Virginia's Warblers." *Condor,* vol. 78, no. 3 (1976), pp. 412–414.

Bull, John L. *Birds of New York State.* New York: Doubleday, 1974.
———. "Wintering Tennessee Warblers." *The Auk,* vol. 78, no. 2 (1961), pp. 263–264.
Bull, John, and John Farrand, Jr. *The Audubon Society Field Guide to North American Birds, Eastern Region.* New York: Alfred A. Knopf, 1977.
Burleigh, Thomas D. *Birds of Idaho.* Caldwell, Idaho: The Caston Printers, 1974.
———. *Georgia Birds.* Norman, Oklahoma: University of Oklahoma Press, 1958.
Burns, Frank L. "The Worm-eating Warbler." *Bird-Lore,* vol. 7 (1905), pp. 137–139.
Burtch, Verdi. "Nesting of the Cerulean Warbler." *Oologist,* vol. 57 (1940), pp. 44–55.
Cadbury, Joseph M. "An Old Warbler." *Bird-Banding,* vol. 40, no. 3 (1969), p. 255.
Campbell, Louis W. "Unusual Nesting Site of the Prothonotary Warbler." *Wilson Bulletin,* vol. 42, no. 4 (1930), p. 292.
Carter, T. Donald. "Six Years with a Brewster's Warbler." *The Auk,* vol. 61, no. 1 (1944), pp. 48–61.
Chamberlain, Montague. "The Nesting Habits of the Cape May Warbler (*Dendroica tigrina*)." *The Auk,* vol. 2 (1885), pp. 33–36.
Chapman, Frank M. *Handbook of Birds of Eastern North America.* New York: D. Appleton-Century Company, 1912.
———. *The Warblers of North America.* 3d ed. New York: D. Appleton and Company, 1917.
Choate, Ernest A. *The Dictionary of American Bird Names.* Boston: Gambit, 1973.
Cockrum, E. Lendell. "A Check-list and Bibliography of Hybrid Birds in North American North of Mexico." *Wilson Bulletin,* vol. 64, no. 3 (1952), pp. 140–159.
Confer, John L., and Kristine Knapp. "The Changing Proportions of Blue-winged and Golden-winged Warblers in Tompkins County and Their Habitat Selection." *Kingbird,* vol. 29, no. 1 (1979), pp. 8–14.
———. "Golden-winged Warblers and Blue-winged Warblers: The Relative Success of a Habitat Specialist and a Generalist." *The Auk,* vol. 98, no. 1 (1981), pp. 108–114.
Coues, Elliott. *Key to North American Birds.* 5th ed. Boston: Dana Estes and Company, 1903.
Cox, George W. "Hybridization Between Mourning and MacGillivray's Warblers." *The Auk,* vol. 90, no. 1 (1973), pp. 190–191.
Cruickshank, Allan D. "Nesting Heights of Some Woodland Warblers in Maine." *Wilson Bulletin,* vol. 68, no. 2 (1956), p. 157.
Dawson, William Leon. *Birds of California,* vol. 1. San Diego: South Moulton Company, 1924.
Decker, F. R., and J. H. Bowles, "Bird Notes from Chelan County, Washington." *Murrelet,* vol. 4 (1923), p. 16.
DeGraaf, Richard M.; and Gretchin M. Witman, John W. Lanier, Barbara J. Hill, and James M. Keniston. *Forest Habitat for Birds of the Northeast.* Washington, D.C.: Forest Service, United States Department of Agriculture, 1980.

Dennis, John V. "Fall Departure of the Yellow-breasted Chat (*Icteria virens*) in Eastern North America." *Bird-Banding,* vol. 38, no. 2 (1967), pp. 130–135.

———. "Some Aspects of the Breeding Ecology of the Yellow-breasted Chat (*Icteria virens*) in Eastern North America." *Bird-Banding,* vol. 29, no. 3 (1958), pp. 169–183.

Dickey, Samuel S. "The Worm-eating Warbler." *Cardinal,* vol. 3 (1934), pp. 179–184.

Dietrich, E. J. "Some Notes on MacGillivray's Warbler." *Oologist,* vol. 31 (1914), pp. 105–111.

Durant, Mary, and Michael Harwood. *On the Road with John James Audubon.* New York: Dodd, Mead and Company, 1980.

Eaton, Stephen W. "A Life History Study of the Louisiana Waterthrush." *Wilson Bulletin,* vol. 70, no. 3 (1958), pp. 211–236.

———. *A Life History Study of Seirus noveboracensis.* Vol. XIX. *Science Studies.* St. Bonaventure, New York: St. Bonaventure University, 1957.

———. "Wood Warblers Wintering in Cuba." *Wilson Bulletin,* vol. 65, no. 3 (1953), pp. 169–174.

Emlen, John T. "Interactions of Migrants and Resident Land Birds in Florida and Bahama Pinelands." In *Migrant Birds in the Neotropics,* edited by Allen Keast and Eugene S. Morton. Washington, D.C.: Smithsonian Institution Press, 1980.

———. "Territorial Aggression in Wintering Warblers at Bahama Agave Blossoms." *Wilson Bulletin,* vol. 85, no. 1 (1973), pp. 71–74.

Faaborg, John. "A Bleak Future for Missouri's Warblers." *Missouri Conservation,* (May 1981), pp. 29–31.

Ficken, Millicent S. "Agnostic Behavior and Territory in the American Redstart." *The Auk,* vol. 79, no. 4 (1962), pp. 607–632.

———. "Courtship of the American Redstart." *The Auk,* vol. 80, no. 3 (1963), pp. 307–317.

Ficken, Millicent S., and Robert W. Ficken. "Age-specific Differences in the Breeding Behavior and Ecology of the American Redstart." *Wilson Bulletin,* vol. 79, no. 2 (1967), pp. 188–199.

———. "Comparative Ethology of the Chestnut-sided Warbler, Yellow Warbler, and American Redstart." *Wilson Bulletin,* vol. 77, no. 4 (1965), pp. 363–375.

———. "The Comparative Ethology of the Wood Warblers: A Review." *Living Bird,* First Annual (1962), pp. 103–122.

———. "Is the Golden-winged Warbler a Social Mimic of the Black-capped Chickadee?" *Wilson Bulletin,* vol. 86 (1974), pp. 468–471.

———. "Some Aberrant Characters of the Yellow-breasted Chat, *Icteria virens.*" *The Auk,* vol. 79, no. 4 (1962), pp. 718–719.

Fisher, Allen C., Jr. "Mysteries of Bird Migration." *National Geographic Magazine,* vol. 156, no. 2 (1979), pp. 154–193.

Forbush, Edward Howe, and John Bichard May. *Natural History of the Birds of Eastern and Central North America.* Boston: Houghton Mifflin Company, 1939.

Friedmann, Herbert. *Additional Data on the Host Relations of the Parasitic Cowbirds.* Washington, D.C.: Smithsonian Miscellaneous Collections, vol. 149, no. 11, 1966.

———. *The Cowbirds.* Springfield, Illinois, and Baltimore, Maryland: Charles C. Thomas, 1929.

———. *Host Relations of the Parasitic Cowbirds.* Washington, D.C.: United States National Museum Bulletin no. 233, 1963.

Friedmann, Herbert, Lloyd F. Kiff, and Stephen I. Rothstein. *A Further Contribution to the Knowledge of the Host Relations of the Parasitic Cowbirds.* Washington, D.C.: Smithsonian Contributions to Zoology, no. 235, 1977.

Gabrielson, Ira N., and Stanley G. Jewett. *Birds of Oregon.* Corvallis, Oregon: Oregon State College, 1940.

George, William G. *The Classification of the Olive Warbler (Peucedramus taeniatus).* American Museum Novitates, no. 2103. New York: American Museum of Natural History, 1962.

Gill, Frank B. "Historical Aspects of Hybridization Between Blue-winged and Golden-winged Warblers." *The Auk,* vol. 97, no. 1 (1980), pp. 1–18.

Gilman, M. French. "Nesting Notes on Lucy's Warbler." *Condor,* vol. 11 (1909), pp. 166–168.

Graham, Frank, Jr. "For Migrants, No Winter Home?" *Audubon,* vol. 82, no. 6 (1980), pp. 14–16.

Grimes, Samuel A. "Injury Feigning by Birds." *The Auk,* vol. 53, no. 4 (1936), pp. 478–480.

Griscom, Ludlow. "Common Sense in Common Names." *Wilson Bulletin,* vol. 59, no. 3 (1947), pp. 131–138.

Griscom, Ludlow, and Alexander Sprunt, Jr. *The Warblers of America.* New York: The Devin-Adair Company, 1957. Also rev. ed., 1979.

Gruson, Edward S. *Words for Birds.* New York: Quadrangle Books, 1972.

Hall, George A. "Hybridization Between Mourning and MacGillivray's Warblers." *Bird-Banding,* vol. 50, no. 2 (1979), pp. 101–107.

Haller, Karl W. "A New Wood Warbler for West Virginia." *Cardinal,* vol. 5 (1940), pp. 49–53.

Hann, Harry W. "Life History of the Oven-bird in Southern Michigan." *Wilson Bulletin,* vol. 49 (1937), pp. 146–237.

Harding, Katherine C. "Nesting Habits of the Black-throated Blue Warbler." *The Auk,* vol. 48, no. 4 (1931), 512–522.

Harrison, George H. *Roger Tory Peterson's Dozen Birding Hot Spots.* New York: Simon & Schuster, 1976.

Harrison, Hal H. *American Birds in Color.* New York: William Wise Company, 1955.

———. *A Field Guide to Birds' Nests.* Boston: Houghton Mifflin Company, 1975.

———. *A Field Guide to Western Birds' Nests.* Boston: Houghton Mifflin Company, 1979.

———. "Notes and Observations on the Wilson's Warbler." *Wilson Bulletin,* vol. 63, no. 3 (1951), pp. 143–148.

———. "A Study in Nesting Diversity." *Living Bird Quarterly,* vol. 2, no. 3 (1983), pp. 26–29.

Harwood, Michael. "Kirtland's Warbler . . . A Born Loser?" *Audubon,* vol. 83, no. 3 (1981), pp. 98–111.

Hausman, Leon A. "On the Winter Food of the Tree Swallow (*Iridoprocne bico-*

lor) and the Myrtle Warbler (*Dendroica coronata*)." *American Naturalist,* vol. 61 (1927), pp. 379–382.

Hiatt, Robert W. "A Singing Female Ovenbird." *Condor,* vol. 45, no. 1 (1943), p. 158.

Hofslund, P. B. "The Genus *Oporornis.*" *Flicker,* vol. 34, no. 2 (1962), pp. 43–47.

———. "A Life History of the Yellowthroat, *Geothlypis trichas.*" *Proceedings of Minnesota Academy of Science,* vol. 27, (1960), pp. 144–174.

Howell, Arthur H. *Birds of Alabama.* 2d ed. Montgomery, Alabama: Department of Game and Fisheries of Alabama, 1928.

———. *Florida Bird Life.* New York: Coward-McCann, Inc., 1932.

Hubbard, John P. "Geographic Variation in the *Dendroica coronata* Complex." *Wilson Bulletin,* vol. 82, no. 4 (1970), pp. 355–369.

———. "The Relationships and Evolution of the *Dendroica coronata* Complex." *The Auk,* vol. 86, no. 3 (1969), pp. 393–432.

Huff, N. L. "The Nest and Habits of the Connecticut Warbler in Minnesota." *The Auk,* vol. 46, no. 4 (1929), pp. 455–465.

Imhof, Thomas A. *Alabama Birds.* Tuscaloosa, Alabama: University of Alabama Press, 1962.

Jaeger, Edmund C. *The Biologist's Handbook of Pronunciations.* Springfield, Illinois: Charles C. Thomas, 1960.

James, Pauline. "Destruction of Warblers on Padre Island, Texas, in May, 1951." *Wilson Bulletin,* vol. 68, no. 3 (1956), pp. 224–227.

Jewett, Stanley G. "Hybridization of Hermit and Townsend Warblers." *Condor,* vol. 46 no. 1 (1944), pp. 23–24.

Jewett, Stanley G., Walter P. Taylor, William T. Shaw, and John W. Aldrich. *Birds of Washington State.* Seattle: University of Washington Press, 1953.

Johnson, Ned K. "Breeding Distribution of Nashville and Virginia's Warblers." *The Auk,* vol. 93, no. 2 (1976), pp. 219–230.

Jones, Lynds. "Warbler Songs." *Wilson Bulletin,* vol. 12, no. 1 (1900), pp. 1–57.

Kammeraad, Jack W. "Notes on Nesting and Survival of Yellow Warblers." *Jack-Pine Warbler,* vol. 44, no. 3 (1966), pp. 124–129.

Keast, Allen, and Eugene S. Morton, Eds. *Migrant Birds in the Neotropics: Ecology, Behavior, Distribution, and Conservation.* Washington, D.C.: Smithsonian Institution Press, 1980.

Kendeigh, S. Charles. *Bird Population Studies in the Coniferous Forest Biome During a Spruce Budworm Outbreak.* Ontario, Canada: Department of Land and Forests, Biological Bulletin no. 1, 1947.

———. "Nesting Behavior of Wood Warblers." *Wilson Bulletin,* vol. 57, no. 3 (1945), pp. 145–164.

Kennard, John H. "Longevity Records of North American Birds." *Bird-Banding,* vol. 46, no. 1 (1975), pp. 55–73.

Knight, Ora Willis. *The Birds of Maine.* Bangor, Maine: Charles H. Glass and Company, 1908.

———. "Contributions to the Life History of the Yellow Palm Warbler." *Journal of Maine Ornithological Society,* vol. 6, no. 2 (1904), pp. 36–41.

Krause, Herbert. "Nesting of a Pair of Canada Warblers." *Living Bird,* 4th Annual, (1965), pp. 5–11.

Kunkle, Donald E. "Unusual Feeding Tactic by a Migrating Myrtle Warbler." *Wilson Bulletin,* vol. 75, no. 1 (1963), p. 89.

Lanyon, Wesley E., and John Bull. "Identification of Connecticut, Mourning and MacGillivray's Warblers." *Bird-Banding,* vol. 38, no. 3 (1967), pp. 187–194.

Lawrence, Louise de Kiriline. "Comparative Study of the Nesting Behavior of Chestnut-sided and Nashville Warblers." *The Auk,* vol. 65, no. 2 (1948), pp. 204–219.

———. "Notes on the Nesting Behavior of the Blackburnian Warbler." *Wilson Bulletin,* vol. 65, no. 3 (1953), pp. 135–144.

Leberman, Robert C. *The Birds of the Ligonier Valley.* Pittsburgh: Carnegie Museum of Natural History, Special Publication no. 3, 1976.

Lein, M. Ross. "Display Behavior of Ovenbirds (*Seiurus aurocapillus*)." *Wilson Bulletin,* vol. 92, no. 3 (1980), pp. 312–329.

Lincoln, Frederick H., Jr.; rev. ed. by Steven R. Peterson. *The Migration of Birds.* Washington, D.C.: U.S. Department of Interior Circular 16, 1935.

Long, Ralph H. *Native Birds of Mount Desert Island and Acadia National Park.* Southwest Harbor, Maine: 1982.

Lowery, George H., Jr. *Louisiana Birds.* 3d ed. Baton Rouge: Louisiana State University Press, 1974.

Maciula, Stanley J. "Worm-eating Warbler 'Adopts' Ovenbird Nestlings." *The Auk,* vol. 77 no. 2 (1960), p. 220.

Marshall, Judy, and Russell P. Balda. "The Breeding Ecology of the Painted Redstart." *Condor,* vol. 76, no. 1 (1974), pp. 89–101.

Marvel, C. S. "Unusual Feeding Behavior of a Cape May Warbler." *The Auk,* vol. 65, no. 4 (1948), p. 599.

Mason, C. Russell. "Cape May Warbler in Middle America." *The Auk,* vol. 93, no. 1 (1976), pp. 167–169.

Mayfield, Harold. *The Kirtland's Warbler.* Bloomfield Hills, Michigan: Cranbrook Institute of Science, 1960.

———. "The Numbers of Kirtland's Warblers." *Jack-Pine Warbler,* vol. 53, no. 2 (1975), pp. 39–47.

———. "Winter Habitat of Kirtland's Warbler." *Wilson Bulletin,* vol. 84 (1972), pp. 347–349.

Maynard, Charles Johnson. *The Warblers of New England.* West Newton, Massachusetts: C. J. Maynard, 1901.

McNicholl, Martin K., and J. Paul Goossen. "Warblers Feeding from Ice." *Wilson Bulletin,* vol. 92, no. 1 (1980), p. 121.

Meanley, Brooke. *Natural History of Swainson's Warbler.* Washington, D.C.: U.S. Department of the Interior, North American Fauna no. 69, 1971.

———. "Some Observations on Habitats of the Swainson's Warbler." *Living Bird,* 5th Annual (1966), pp. 151–165.

Mendall, Howard L. "Nesting of the Bay-breasted Warbler." *The Auk,* vol. 54, no. 4 (1937), pp. 429–439.

Mengel, Robert M. *The Birds of Kentucky.* Ornithological Monographs no. 3, American Ornithologists' Union, 1965.

———. "The Probable History of Species Formation in Some Northern Wood Warblers (Parulidae)." *Living Bird,* 3d Annual (1964), pp. 9–43.

Meyer, Henry, and Ruth Reed Nevius. "Some Observations on the Nesting and Development of the Prothonotary Warbler." *Migrant,* vol. 14 (June 1943), pp. 31–36.

Morrison, Michael L. "The Structure of Western Warbler Assemblages: Ecomorphological Analysis of the Black-throated Gray and Hermit Warblers." *The Auk,* vol. 99, no. 3 (1982), pp. 503–513.

Morse, Douglass H. "Foraging and Coexistence of Spruce-Woods Warblers." *Living Bird,* 18th Annual (1979–80), pp. 7–25.

———. "Foraging of Pine Warblers Allopatric and Sympatric to Yellow-throated Warblers." *Wilson Bulletin,* vol. 86, no. 4 (1974), pp. 474–477.

———. "Habitat Use by the Blackpoll Warbler." *Wilson Bulletin,* vol. 91, no. 2 (1979), pp. 234–243.

———. "Populations of Bay-breasted and Cape May Warblers During an Outbreak of the Spruce Budworm." *Wilson Bulletin,* vol. 90, no. 3 (1978), pp. 404–413.

Mosher, Franklin Herbert. *Food Plants of the Gypsy Moth in America,* U.S. Department of Agriculture Bulletin no. 250, 1915.

Mousley, Henry. "A Further Study of the Home Life of the Northern Parula and of the Yellow Warbler and Ovenbird." *The Auk,* vol. 43 (1926), pp. 184–197.

———. "A Study of the Home Life of the Northern Parula and Other Warblers at Hatley, Stanstead County, Quebec, 1921–1922." *The Auk,* vol. 41, no. 2 (1924), pp. 263–288.

Nice, Margaret Morse. A" Study of a Nesting Magnolia Warbler (*Dendroica magnolia*)." *Wilson Bulletin,* vol. 38 (1926), pp. 185–199.

Nice, Margaret Morse, and Leonard B. Nice. "A Study of Two Nests of the Black-throated Green Warbler." *Bird-Banding,* vol. 3, 1932.

Nolan, Val, Jr. *Ecology and Behavior of the Prairie Warbler* (*Dendroica discolor*). Washington, D.C.: Ornithological Monographs no. 26, American Ornithologists' Union, 1978.

Oberholser, Harry C. *The Bird Life of Texas,* vol. 2. Austin: University of Texas Press, 1974.

Ogburn, Charlton. *The Adventure of Birds.* New York: William Morrow and Company, Inc., 1976.

Owre, Oscar T. "Predation by the Chuck-will's-widow upon Migrating Warblers." *Wilson Bulletin,* vol. 79, no. 3 (1967), p. 342.

Palmer, Ralph S. *Maine Birds.* Cambridge, Massachusetts: Bulletin of the Museum of Comparative Zoology at Harvard College, vol. 102, 1949.

Palmer, T. S. "Notes on Persons Whose Names Appear in the Nomenclature of California Birds." Condor, vol. 30, no. 5 (1928), pp. 261–307.

Parkes, Kenneth C. "The Genetics of the Golden-winged × Blue-winged Warbler Complex." *Wilson Bulletin,* vol. 63, no. 1 (1951), pp. 5–15.

———. "Still Another Parulid Intergeneric Hybrid and Its Taxonomic and Evolutionary Implications." *The Auk,* vol. 95, no. 4 (1978), pp. 682–690.

———. "Taxonomic Relationships Among the American Redstarts." *Wilson Bulletin,* vol. 73, no. 4 (1961), pp. 374–379.

Parmelee, David F. "The Nest of the Northern Parula." *Living Bird,* 12th Annual (1973), pp. 197–199.

Pasquier, Roger F. *Watching Birds.* Boston: Houghton Mifflin Company, 1977.
Pasquier, Roger F., and Eugene S. Morton. "For Avian Migrants a Tropical Vacation Is Not a Bed of Roses." *Smithsonian,* vol. 13, no. 7 (1982), pp. 169–178.
Pearson, T. Gilbert. *Birds of America.* New York: Garden City Publishing Company, 1936.
Pearson, T. Gilbert, C. S. Brimley and H. H. Brimley. *Birds of North Carolina.* Revised by David L. Wray and Harry T. Davis. Raleigh: North Carolina Department of Agriculture, 1959.
Peterson, Roger Tory. "A Bird by Any Other Name." *Audubon* magazine, vol. 44, no. 5 (1942), pp. 277–280.
———. *A Field Guide to the Birds.* 4th ed. Boston: Houghton Mifflin Company, 1980.
———. *A Field Guide to Western Birds.* 2d ed. Boston: Houghton Mifflin Company, 1961.
Peterson, Roger Tory, and Edward L. Chalif. *A Field Guide to Mexican Birds.* Boston: Houghton Mifflin Company, 1973.
Petrides, George A. "A Life History Study of the Yellow-breasted Chat." *Wilson Bulletin,* vol. 50, no. 3 (1938), pp. 184–189.
Pettingill, Olin Sewall, Jr. *Ornithology in Laboratory and Field.* 4th ed. Minneapolis: Burgess Publishing Company, 1970.
Phillips, Allan R. "The Races of MacGillivray's Warbler." *The Auk,* vol. 64, no. 2 (1947), pp. 296–300.
Phillips, Allan, Joe Marshall, and Gale Monson. *The Birds of Arizona.* Tucson: The University of Arizona Press, 1964.
Phillips, Allan R., and J. Dan Webster. "Grace's Warbler in Mexico." *The Auk,* vol. 78, no. 4 (1961), pp. 551–553.
Pitelka, Frank A. "Breeding Behavior of the Black-throated Green Warbler." *Wilson Bulletin,* vol. 52, no. 1 (1940), pp. 3–18.
Porter, Eliot. *Birds of North America, a Personal Selection.* New York: E. P. Dutton and Company, 1972.
Powell, George V. N. and H. Lee Jones, "An Observation of Polygyny in the Common Yellowthroat." *Wilson Bulletin,* vol. 96, no. 4 (1978), p. 656.
Pulich, Warren M. *The Golden-cheeked Warbler.* Austin: Texas Parks and Wildlife Department, 1976.
Radabaugh, Bruce E. "Kirtland's Warbler and Its Bahama Wintering Grounds." *Wilson Bulletin,* vol. 86, no. 4 (1974), pp. 374–383.
———. "Polygamy in the Kirtland's Warbler." *Jack-Pine Warbler,* vol. 50, no. 2 (1972), pp. 48–52.
Rand, Austin L. "Bass Eats Yellowthroat, Young Stilts, and Young Ducks." *The Auk,* vol. 60, no. 1 (1943), p. 95.
Rea, Gene. "Black and White Warbler Feeding Young of Worm-eating Warbler." *Wilson Bulletin,* vol. 57, no. 4 (1945), p. 262.
Robbins, Chandler S., Bertel Bruun, and Herbert S. Zim. *A Guide to Field Identification of Birds of North America.* New York: Golden Press, 1966.
Roberts, T. S. *The Birds of Minnesota.* Minneapolis: University of Minnesota Press, 1936.
Ross, Lucretius H. "Northern Yellow-throat, *Geothlypis trichas,* Caught in a Spider Web." *The Auk,* vol. 67, no. 4 (1950), pp. 521–522.

Ryel, Lawrence A. "The Kirtland's Warbler in 1982." *Jack-Pine Warbler,* vol. 60, no. 4 (1982), pp. 147–150.

———. "Population Change in the Kirtland's Warbler." *Jack-Pine Warbler,* vol. 59, no. 3 (1981), pp. 76–91.

Salt, W. Ray. *Alberta Vireos and Wood Warblers.* Alberta: Provincial Museum and Archives of Alberta Publication no. 3, 1973.

Saunders, Aretas A. *A Guide to Bird Songs.* New York: Doubleday and Company, Inc., 1951.

Saunders, W. E. "Nesting Habits of the Cerulean Warbler." *The Auk,* vol. 17, (Oct. 1900), pp. 358–362.

Schrantz, F. G. "Nest Life of the Eastern Yellow Warbler." *The Auk,* vol. 60, no. 3 (1943), pp. 367–387.

Schwartz, Paul. "Some Considerations on Migratory Birds." In *Migrant Birds in the Neotropics,* edited by Allen Keast and Eugene S. Morton. Washington, D.C.: Smithsonian Institution Press, 1980.

Seton, Ernest E. T. "Nest and Habits of the Connecticut Warbler (*Oporornis agilis*)." *The Auk,* vol. 1, (April 1884), pp. 192–193.

Shake, William F., and James P. Mattsson. "Three Years of Cowbird Control: An Effort to Save the Kirtland's Warbler." *Jack-Pine Warbler,* vol. 53, no. 2 (1975), pp. 48–53.

Short, L. L. "The Blue-winged Warbler and Golden-winged Warbler in Central New York State." *Kingbird,* vol. 12 (1962), pp. 59–67.

———. "Hybridization of Wood Warblers: *Vermivora pinus* and *V. chrysoptera.*" Proceedings of the 13th International Ornithological Congress, pp. 147–160, 1963.

Shuler, Jay. "Bachman's Phantom Warbler." *Birding,* vol. 9, no. 6 (1977). pp. 245–250.

———. "Bachman's Warbler Habitat." *Chat,* vol. 41, no. 2 (1977), pp. 19–23.

———. "Clutch Size and Onset of Laying in Bachman's Warbler." *Chat,* vol. 43, no. 2 (1979), pp. 27–29.

Sibley, Charles G., and Jon B. Ahlquist. "The Relationship of the Yellow-breasted Chat (*Icteria virens*) and the Alleged Slowdown in the Rate of Macromolecular Evolution in Birds." *Postilla,* no. 187, 1982.

Skaggs, Merit B. "A Study of the Prothonotary Warbler in Northern Ohio." *Redstart,* vol. 16, no. 4 (1949), pp. 56–63.

Skutch, Alexander F. "Family Parulidae." In *Life Histories of Central American Birds,* pp. 339–386. Cooper Ornithological Society, Pacific Coast Avifauna no. 31, 1954.

———. *Life Histories of Central American Highland Birds.* Publications of the Nuttall Ornithological Club, no. 7, 1967.

Smith, Wendell P. "Observations of the Nesting Habits of the Black and White Warbler." *Bird-Banding,* vol. 5 (Jan. 1934), pp. 31–35.

Snyder, Dorothy E. "A Recent Colima Warbler's Nest." *The Auk,* vol. 74, no. 1 (1957), pp. 97–98.

Sprunt, Alexander, Jr. *Florida Bird Life.* New York: Coward-McCann, 1954.

Sprunt, Alexander, Jr., and E. Burnham Chamberlain. *South Carolina Bird Life.* Rev. ed. Columbia: University of South Carolina Press, 1970.

Squires, W. Austin. *The Birds of New Brunswick.* Saint John, New Brunswick: Monographic Series, no. 7, New Brunswick Museum, 1976.

Stanwood, Cornelia J. "A Series of Nests of the Magnolia Warbler." *The Auk,* vol. 27 (1910), pp. 384–389.

Stein, Robert C. "A Comparative Study of Songs Recorded from Five Closely Related Warblers." *Living Bird,* 1st Annual (1962), pp. 61–71.

Stevens, O. A. "The First Descriptions of North American Birds." *Wilson Bulletin,* vol. 43, no. 3 (1936), pp. 203–215.

Stevenson, Henry M. "The Recent History of Bachman's Warbler." *Wilson Bulletin,* vol. 84, no. 3 (1972), pp. 344–347.

Stewart, Robert E. "A Life History Story of the Yellow-throat." *Wilson Bulletin,* vol. 65, no. 2 (1953), pp. 99–115.

Stewart, Robert E., and John W. Aldrich. "Ecological Studies of Breeding Bird Populations in Northern Maine." *Ecology,* vol. 33, no. 2 (1952), pp. 226–238.

Stewart, Robert M., R. Phillip Henderson, and Kate Darling. "Breeding Ecology of the Wilson's Warbler in the High Sierra Nevada, California." *Living Bird,* 16th Annual (1977), pp. 83–102.

Street, Phillips B. *Birds of the Pocono Mountains, Pennsylvania.* Philadelphia: Delaware Valley Ornithological Club, 1956.

Sturm, Louis. "A Study of the Nesting Activities of the American Redstart." *The Auk,* vol. 62, no. 2 (1945), pp. 189–206.

Sutton, George Miksch. "An Expedition to the Big Bend Country." *Cardinal,* vol. 4, no. 1 (1935), pp. 1–7.

Swain, J. Morton. "Contributions to the Life History of Wilson's Warbler." *Journal Maine Ornithological Society,* vol. 6, no. 3 (1904), pp. 59–62.

Sykes, Paul W., Jr. "The Eightieth Audubon Christmas Bird Count, Florida." *American Birds,* vol. 34, no. 4 (1980), p. 340.

Tate, James, Jr. "Nesting and Development of the Chestnut-sided Warbler." *Jack-Pine Warbler,* vol. 48, no. 2 (1970), pp. 57–65.

Taverner, P. A. *Birds of Western Canada.* National Museum of Canada Bulletin no. 41, 1928.

Taylor, Walter Kingsley. "Migration of the Common Yellow-throat with an Emphasis on Florida." *Bird-Banding,* vol. 47, no. 4 (1976), pp. 319–332.

Taylor, Walter Kingsley, and Bruce H. Anderson. "Nocturnal Migrants Killed at a Central Florida TV Tower, Autumns 1969–71." *Wilson Bulletin,* vol. 85, no. 1 (1973), pp. 42–51.

Terborgh, John W. "The Conservation Status of Neotropical Migrants: Present and Future." In *Migrant Birds in the Neotropics,* edited by Allen Keast and Eugene S. Morton. Smithsonian Institution Press, 1980, pp. 21–30.

Terres, John K. *The Audubon Society Encyclopedia of North American Birds.* New York: Alfred A. Knopf, 1980.

Todd, W. E. Clyde. *Birds of Western Pennsylvania.* Pittsburgh: University of Pittsburgh Press, 1940.

Tufts, Robie W. *The Birds of Nova Scotia.* Halifax: Nova Scotia Museum, 1973.

Tyler, Winsor Marrett. "Black-and-white Warbler." In *Life History of North American Wood Warblers,* edited by Arthur Cleveland Bent, Smithsonian Institution Bulletin no. 203, 1953.

Vaiden, M. G. "The Prothonotary Warbler." *Oologist,* vol. 57, no. 4 (1940), p. 43.

Van Tyne, Josselyn. *The Discovery of the Nest of the Colima Warbler (Vermivora crissalis).* Ann Arbor: University of Michigan Miscellaneous Publications no. 33, 1936.

Van Tyne, Josselyn, and George Miksch Sutton. *The Birds of Brewster County, Texas.* Ann Arbor: University of Michigan Miscellaneous Publications no. 37, 1937.

Verner, Jared, and Mary F. Willson. *Mating Systems, Sexual Dimorphism, and the Role of Male North American Passerine Birds in the Nesting Cycle,* Ornithological Monographs no. 9, American Ornithologists' Union, 1969.

Walkinshaw, Lawrence H. "Kirtland's Warbler—Endangered." *American Birds,* vol. 26, no. 1 (1972), pp. 3–9.

———. "Nesting Studies of the Prothonotary Warbler." *Bird-Banding,* vol. 9 (1938), pp. 32–46.

———. "Observations on Summering and Migrating Warblers in Muskegon County, Michigan." *Jack-Pine Warbler,* vol. 46, no. 2 (1968), pp. 42–56.

———. "The Prairie Warbler in Michigan." *Jack-Pine Warbler,* vol. 37 (1959), pp. 54–63.

———"The Prothonotary Warbler: A Comparison of Nesting Conditions in Tennessee and Michigan." *Wilson Bulletin,* vol. 53, no. 1 (1941), pp. 3–21.

Walkinshaw, Lawrence H., and Mark A. Wolf. "Distribution of the Palm Warbler and Its Status in Michigan." *Wilson Bulletin,* vol. 9, no. 4 (1957), pp. 338–351.

Wauer, Roland H. *Birds of Big Bend National Park and Vicinity.* Austin: University of Texas Press, 1973.

Webster, J. Dan. "A Revision of Grace's Warbler." *The Auk,* vol. 78, no. 4 (1961), pp. 554–566.

Welsh, Daniel A. "Breeding and Territoriality of the Palm Warbler in a Nova Scotia Bog." *Canadian Field Naturalist,* vol. 85, no. 1 (1971), pp. 31–37.

Wetmore, Alexander. "Observations on the Habits of Birds at Lake Burford, New Mexico." *The Auk,* vol. 37 (1920), pp. 393–412.

———. *Song and Garden Birds of North America.* Washington, D.C.: National Geographic Society, 1964.

Widmann, Otto. "The Summer Home of Bachman's Warbler No Longer Unknown: A Common Breeder in the St. Francis River Region of Southeastern Missouri and Northeastern Arkansas." *The Auk,* vol. 14 (July 1897), pp. 305–310.

Willard, Francis Cottle. "The Olive Warbler (*Dendroica olivacea*) in Southern Arizona." *Condor,* vol. 12 (1910), pp. 104–107.

Wood, Norman A. *The Birds of Michigan.* Ann Arbor: Miscellaneous Publications, Museum of Zoology, University of Michigan, no. 75, 1951.

Wright, Horace W. "The Orange-crowned Warbler as a Fall and Winter Visitant in the Region of Boston, Massachusetts." *The Auk,* vol. 34 no. 1 (1917), pp. 11–27.

Index

PICTURE CREDITS

(Continued from copyright page)

All photographs were taken by the author except the following:

COLOR PHOTOGRAPHS

Herbert Clarke: Plate 2 (bottom), Plate 24 (bottom).
Betty Darling Cottrille: Plate 3 (bottom), Plate 17, Plate 22 (bottom left and right).
John H. Dick: Plate 2 (top left).
Bill Dyer: Plate 1, Plate 6 (bottom), Plate 12 (top left), Plate 18 (center), Plate 20 (top right), Plate 24 (center).
John H. Hoffman: Plate 10 (bottom left).

BLACK-AND-WHITE PHOTOGRAPHS

Emmet R. Blake: p. 76.
Bob and Elsie Boggs: pp. 60, 137 (top).
Herbert Clarke: p. 140 (bottom).
Betty Darling Cottrille: pp. 44, 45, 89, 114 (bottom), 120, 124, 197, 198 (top), 208, 214 (top), 257, 264, 271, 284.
Bill Dyer: pp. 32, 50 (bottom), 54 (bottom), 103, 119, 156, 175, 176, 177, 196 (top), 198 (bottom), 258.
Samuel A. Grimes: pp. 85, 220, 231 (bottom), 232 (bottom).
Ruth Grom: p. 29.
E. M. Hall: p. 62 (bottom).
George H. Harrison: pp. 19, 24, 28, 31, 158, 161, 173, 188.
John H. Hoffman: pp. 61, 137 (bottom).
N. L. Huff (courtesy Walter J. Breckenridge): pp. 256, 259.
Leigh Yawkey Woodson Art Museum: p. 162.
Ray Quigley: pp. 37, 62 (top).
Ron Quigley: p. 140 (top).
Jay Shuler: p. 300.
Smithsonian Institution: pp. 71, 114 (top), 141, 196 (bottom), 310.
U.S. Department of the Interior: p. 75.
Ken Vierling: pp. 52, 179, 252 (bottom).

ABOUT THE AUTHOR

Hal H. Harrison was born in Tarentum, Pennsylvania, in 1906. Still a vigorous writer, photographer and lecturer, he has been published in scores of magazines and newspapers, and is the author of eight previous books, including two Peterson Field Guides, *A Field Guide to Birds' Nests* and *A Field Guide to Western Birds' Nests.* He has studied Wood Warblers for more than 30 years in their summer homes and during migrations north and south (and, in many cases, in their winter homes as well).

Hal Harrison lives with his wife, Mada, in Fort Lauderdale, Florida, and Southwest Harbor, Maine.

The Harrisons' son, George, is also a Simon and Schuster author (*The Backyard Bird Watcher* and *America's Favorite Backyard Birds*).